# THE DIVERSE FACES OF *BACILLUS CEREUS*

# DEDICATION

To my loves, Giorgia and Alesia; to my parents, Taziana and Francesco; to my grandmother, Ester, my grandfather, Gigino, my aunt, Elda, who are today living in the Light; to my Angels, on this earth, over the clouds.

# THE DIVERSE FACES OF
# *BACILLUS CEREUS*

Edited by

**VINCENZO SAVINI**

Amsterdam • Boston • Heidelberg • London
New York • Oxford • Paris • San Diego
San Francisco • Singapore • Sydney • Tokyo
Academic Press is an imprint of Elsevier

Academic Press is an imprint of Elsevier
125 London Wall, London EC2Y 5AS, UK
525 B Street, Suite 1800, San Diego, CA 92101-4495, USA
50 Hampshire Street, 5th Floor, Cambridge, MA 02139, USA
The Boulevard, Langford Lane, Kidlington, Oxford OX5 1GB, UK

**Notices**
Knowledge and best practice in this field are constantly changing. As new research and
experience broaden our understanding, changes in research methods, professional practices, or
medical treatment may become necessary.

Practitioners and researchers must always rely on their own experience and knowledge in
evaluating and using any information, methods, compounds, or experiments described herein. In
using such information or methods they should be mindful of their own safety and the safety of
others, including parties for whom they have a professional responsibility.

To the fullest extent of the law, neither the Publisher nor the authors, contributors, or editors,
assume any liability for any injury and/or damage to persons or property as a matter of products
liability, negligence or otherwise, or from any use or operation of any methods, products,
instructions, or ideas contained in the material herein.

**British Library Cataloguing-in-Publication Data**
A catalogue record for this book is available from the British Library

**Library of Congress Cataloging-in-Publication Data**
A catalog record for this book is available from the Library of Congress

ISBN: 978-0-12-801474-5

For information on all Academic Press publications
visit our website at https://www.elsevier.com/

*Publisher:* Janice Audet
*Acquisition Editor:* Linda Versteeg-Buschman
*Editorial Project Manager:* Halima Williams
*Production Project Manager:* Lucía Pérez
*Designer:* Greg Harris

Typeset by TNQ Books and Journals
www.tnq.co.in

Picture credit: Images on the cover are by Vincenzo Savini. Main picture has been taken at San
Domenico Lake, Villalago (AQ), Italy. Images of the microorganisms have been taken at the Civc
Hospital of Pescara, Italy, and are explained in the text.

# CONTENTS

# CONTRIBUTORS

**Edoardo Carretto**
Clinical Microbiology Laboratory, IRCCS Arcispedale Santa Maria Nuova,
Reggio Emilia, Italy

**Stefano Colombo**
Mérieux NutriSciences, Lyon, France

**Paolo Fazii**
Clinical Microbiology and Virology, Laboratory of Bacteriology and Mycology,
Civic Hospital of Pescara, Pescara, Italy

**Giovanni Gherardi**
Department of Biomedical Sciences, Campus Bio-Medico University of Rome,
Rome, Italy

**Dallas G. Hoover**
Department of Animal and Food Sciences, University of Delaware, Newark, DE, USA

**Sarah M. Markland**
Department of Animal and Food Sciences, University of Delaware, Newark, DE, USA

**Roberta Marrollo**
Clinical Microbiology and Virology, Laboratory of Bacteriology and Mycology,
Civic Hospital of Pescara, Pescara, Italy

**Daniela Onofrillo**
Pediatric Hematology and Oncology Unit, Department of Hematology, Spirito Santo
Hospital, Pescara, Italy

**Eugenio Pontieri**
Department of Biotechnological and Applied Clinical Sciences, University of L'Aquila, Italy

**Vincenzo Savini**
Clinical Microbiology and Virology, Laboratory of Bacteriology and Mycology,
Civic Hospital of Pescara, Pescara, Italy

**Rosa Visiello**
Clinical Microbiology Laboratory, IRCCS Arcispedale Santa Maria Nuova,
Reggio Emilia, Italy

# CHAPTER 1

# Microbiology of *Bacillus cereus*

## Roberta Marrollo

Clinical Microbiology and Virology, Laboratory of Bacteriology and Mycology, Civic Hospital of Pescara, Pescara, Italy

## SUMMARY

The species *Bacillus cereus* belongs to the so-called *B. cereus* group, which includes Gram-positive, aerobic or facultative, sporulating, rods that are almost ubiquitous in the natural environment. The organism is a cause of food poisoning and severe and potentially lethal nonintestinal infections in humans. When *B. cereus* is isolated from a clinical sample it is usually hard to label it as a pathogen or an innocent contaminant, so isolates must be carefully evaluated case by case instead of being aprioristically dismissed as colonizers. In fact, outside its notoriety in the etiology of food-related enteric tract diseases, this bacterium may be involved in a multitude of clinical pictures, such as anthrax-like pneumonias, fulminant bacteremias, devastating central nervous system diseases, postsurgery wound infections, and primary pathologic processes involving skin structures and mimicking clostridial gas gangrene, that usually complicate traumatic events.

## BACKGROUND

It is both fascinating and of concern to observe the evolving spectrum of human diseases associated with *Bacillus cereus*, which, given its environmental ubiquity, has long been labeled as a mere contaminant when isolated from clinical materials of human origin.[1]

The term "bacillus" means a "small rod," while the word "cereus," translated from Latin, stands for "wax-like." This name in fact relies on the easily recognizable phenotype of *B. cereus* that may be observed under a microscope and, macroscopically, on blood-containing agar media, respectively.[1]

*B. cereus* is an aerobic or facultative, motile, sporulating organism that is almost ubiquitously distributed in the natural environment. It is close, phenotypically as well as genetically (we mean based on 16S rRNA analysis), to several other *Bacillus* species, especially its most famous relative *Bacillus anthracis*[2]; these species are parts of the so-called *B. cereus* group (otherwise

*The Diverse Faces of Bacillus cereus*
ISBN 978-0-12-801474-5
http://dx.doi.org/10.1016/B978-0-12-801474-5.00001-3

named as *B. cereus sensu lato*).[3,4] In fact, it is well known today that such above-mentioned *B. cereus* (*sensu lato*) is not indeed only one bacterium but actually represents a wide group of microorganisms (that is, therefore, actually referred to as to the *B. cereus* group) that includes the eight species *B. cereus* (*sensu stricto*), *B. anthracis*, *Bacillus mycoides*, *Bacillus pseudomycoides*, *Bacillus thuringiensis*, *Bacillus weihenstephanensis*, *Bacillus cytotoxicus*, and *Bacillus toyonensis*.[3,4]

B. cereus group is therefore an informal term indicating a genetically highly homogeneous subdivision within the genus *Bacillus*, comprising the above-mentioned recognized species. The type strain of *B. cereus sensu stricto* is the strain ATCC (American Type Culture Collection) 14579, which was originally cultivated from the air in a cowshed more than one century ago.[5]

The more, and unfortunately, famous and genetically close species we mentioned previously, that is *B. anthracis*, is maybe the most deeply investigated within the *B. cereus* group, and its notoriety relies on the fact that it is known as the worrisome agent of a lethal animal and human disease, which is worldwide called "anthrax"; by virtue of this particular pathogenicity, the organism has in recent years emerged as a biological weapon, too.[6,7]

B. thuringiensis is characterized by the ability to display insecticidal properties and is therefore commercially used as a biocontrol microbial agent with the interesting potential to face crop insects.[7,50]

Discrimination between isolates of the *B. cereus* group is challenging and, based on a mere phenotype observation, we may just affirm that *B. anthracis* can be distinguyished from *B. cereus* based on the absence of motility, while *B. mycoides* and *B. pseudomycoides* can be both differentiated from *B. cereus* on a phenotypical base by virtue of the rhizoidal shape of their colonies along with fatty acid composition.[8,9] Again, to confirm the difficulties in distinguishing *B. cereus* group species, it is known that although *B. anthracis, B. cereus*, and *B. thuringiensis* may be recognized by examining their phenotypes and pathological aspects, genome sequencing has shown their close relatedness, and their 16S rRNA gene sequences have been found to share more than 99% similarity.[1] Also, the *B. cereus* group organism that may be observed on this book's cover (on the reader's left hand) was collected at our bacteriology laboratory at the Pescara Civic Hospital, Italy, and identified as *B. cereus/B. mycoides/B. thuringiensis*, by using matrix-assisted laser desorption ionization–time of flight technology, and as *B. cereus/B. thuringiensis* by virtue of 16S rRNA gene sequencing, no deeper discrimination being possible with these two methodologies. Moreover, it seems that the phylogenetic investigations based on chromosomal markers indeed show that there is no taxonomic evidence that justifies separation of *B. cereus* and *B. thuringiensis*, whereas *B. anthracis* can basically be referred to as a clone of *B. cereus*.[1]

Notably, however, the distinguishing traits between these species are encoded by plasmid genes, rather than chromosomal sequences, which are highly mobile genetic elements, even within the *B. cereus* group.[1] In this context, *B. thuringiensis* possesses plasmids carrying *cry* genes that encode the δ-endotoxins, while *B. anthracis* harbors two large plasmids encoding the two major pathogenic factors of this species, that is, pXO1 (encoding the anthrax toxin complex) and pXO2 (encoding the capsule).[1] The role of plasmids in the expression of virulence within the *B. cereus* group is also documented by the finding that, in *B. cereus*, the genetic determinants of the emetic toxin (cereulide), the *ces* genes, are located on a large plasmid as well.[1] This plasmid was found to be almost exclusively carried by a single monomorphic strain cluster of *B. cereus* (*sensu stricto*), although cereulide-producing isolates have been reported to differ from the mentioned highly homogeneous cluster in both genotypic and phenotypic characteristics.[1] As a confirmation, multilocus sequence typing (MLST) identified cereulide-producing strains that belonged to a phylogenetic cluster outside the major monomorphic emetic one.[1] Moreover, two *B. weihenstephanensis* strains were shown to produce cereulide and carry the *cesB* gene.[1] Emetic *B. cereus* has been proposed to differ from nonemetic strains in phenotype properties including hemolysis, starch hydrolysis, lecithinase activity, and temperature range for growth.[1]

The dilemma in defining which are the species that form the *B. cereus* group is then that the main pathogenic factors that may distinguish *B. cereus* from *B. thuringiensis* and *B. anthracis* do not correlate with phylogeny, when phylogeny studies rely on chromosomal markers. Of interest, an exception is represented by a newly discovered cluster made of only three strains, thus far, that includes the *B. cereus* strain NVH 391/98, responsible for three fatal cases of diarrheal syndrome.[1] MLST and genomic sequencing revealed that this group is far enough from the main *B. cereus* group cluster to warrant the status of novel species, named "*Bacillus cytotoxicus*," as informally proposed for these strains.[1] These three isolates can grow at temperatures 6–8 °C higher than the mesophilic *B. cereus* strains; thus they are thermotolerant representatives of the *B. cereus* group.[1] Conversely, it is known that the species *B. weihenstephanensis* was created to distinguish psychrotolerant *B. cereus* strains from mesophilic ones.[1] Several typing methods suggest that *B. weihenstephanensis* groups in a separate clade within the *B. cereus* group, together with *B. mycoides*. However, it has been shown that psychrotolerant *B. cereus* strains do not always conform to the *B. weihenstephanensis* species criteria, and a genetic group including psychrotolerant *B. cereus* and *B. thuringiensis* strains has been identified that is far from the *B. weihenstephanensis* clade, based on phylogeny. Notably, the temperature tolerance limits for *B. cereus* group strains do correlate with diverse phylogenetic clusters.[1]

Although *B. cereus* was originally labeled as a mesophilic organism that grows between 10 and 50 °C (optimum temperature is 35–40 °C),[10,11] increasing numbers of psychrotolerant strains have been described over the past few decades; this led to the description of a novel, psychrotolerant species within the group, the previously mentioned *B. weihenstephanensis*, which has been known to be able to grow below 7 °C but not at 43 °C.[12] *B. cereus sensu stricto*, in conclusion, includes therefore all strains of the *B. cereus* group that cannot belong to any of the species mentioned other than *B. cereus*. Unfortunately, owing to a lack of identification to a species level somewhere in the scientific articles, there is some confusion throughout the literature concerning the description of organisms as *B. cereus sensu lato* or *B. cereus sensu stricto*; thus we may expect that features and clinical syndromes reported in published works have been alternately linked to the former or the latter. Keeping in mind such a microbiological issue, we will refer to *B. cereus sensu stricto* by using the term *B. cereus* in this volume.

*B. cereus*, then, inhabits nature as a spore former and vegetative cell, while it colonizes human tissues in a vegetative status.[13,14] Spores may bear extreme environmental conditions including heat, drying, freezing, and radiation and behave as the infective agent for this microorganism.[13,14]

Being in close relationship to *B. anthracis*, as said, the spore surface antigens of *B. cereus* share epitopes with those of the former, as shown by cross-agglutination.[15] Spores are of particular concern in the food industry since they can be refractory to pasteurization and γ-radiation; also, their hydrophobic nature makes them adhere to surfaces, thus unfortunately supporting the role of this species as a famous contaminant in the food-processing chain, at diverse and several levels.[16,17]

## ECOLOGY

*B. cereus* may be found in various types of soils and sediments, as well as in dust and plants.[1] Particularly, the natural reservoir for this organism in the environment consists of decaying organic matter, along with fresh, river, and marine waters; fomites; vegetables; and the gut of invertebrates, from which soil and food may become contaminated, leading to colonization of the human enteric tract.[1,18,19] Water contamination, particularly, suggests the possibility that waters themselves may behave as a means by which *B. cereus* is introduced into the food-processing chain.[1]

Nevertheless, spores are passively spread and are therefore found even outside the above-mentioned natural habitats. It is thought that *B. cereus sensu lato* is present in soils as spores that then germinate and grow under certain conditions.[1] Germination of spores in fact occurs when these come into contact with organic matter or are taken in by an insect or an animal.[1] Hence, *B. cereus* shows a saprophytic life cycle in which spores germinate in soil, with consequent production of vegetative bacilli, which can then sporulate, thus continuing the life cycle itself.[1] Defecation by the host, or its death, may release cells and spores into the soil, so vegetative cells sporulate and survive until the following intake by a new host.[18,20] Of further interest, a multicellular phenotype characterized by a filamentous mode of growth was described and proposed to be a means of translocation through the soil.[1] A similar mode of growth has also been found in the intestine of insects.[1]

It has also been proposed that such an existence of diverse morphological modes may represent adaptations to diverse life cycles like the "normal" life cycle as a symbiont or the less common pathogenic life cycle in which rapid growth occurs.[1]

Given its almost ubiquitous distribution in food products (such as milk, rice, and pasta), *B. cereus* is ingested in small amounts of cells, so the organism transiently becomes a part of the human gut microbiota, although it is unknown if isolation of *B. cereus* from stools is related to spore germination or to growth of vegetative cells.[18,21]

The ubiquitous low-level presence of *B. cereus* in feed, foods, and the environment would ensure the organism a transient presence in the mammalian gut, including that of humans.[1]

Genomic data seem to suggest, however, that the metabolic capacity of *B. cereus* type strain ATCC 14579 and *B. anthracis* is more adapted to proteins as a nutrient source rather than carbohydrates and that, moreover, genes on which the establishment within a host depends are conserved in those strains.[1]

Nevertheless, *B. cereus* strain ATCC 14579, if compared to *B. cereus* strain ATCC 10987, is able to metabolize a wider amount of carbohydrates than what was originally thought on the basis of genomic analysis only. This information supports the idea that, in addition to having a full life cycle in soils, where it is abundantly present, and growing in food products, *B. cereus* is also adapted to a host-related lifestyle, as a pathogen or behaving as a part of the gut microbiota.[1]

The possible adaptation of this organism to the animal intestine could represent the basis for its proposed probiotic activity. Such a use cannot, however, be uncritically labeled as safe for humans, as all *B. cereus* strains are known to potentially produce at least one of the toxins related to the diarrheal syndrome. Nonetheless, certain strains that produce negligible quantities of toxin at 37 °C have been approved for the above-mentioned purpose by the European Food Safety Authority (EFSA).[1]

## PHENOTYPE

Organisms of the *B. cereus* group, exclusive of *B. anthracis*, show a variety of morphological forms depending on the medium in which they grow. *B. cereus*, particularly, is a large (1.0–1.2 µm by 3.0–5.0 µm) Gram-positive rod that grows on common agar plates.[1]

In body fluids (such as aspirate from the anterior chamber of the eye) or broth cultures, *B. cereus* presents as straight or slightly curved slender rods, with square ends, and is arranged singly or in short chains. Bacterial cells from agar plates will be characterized instead by a more uniform rod-shaped morphology with, typically, an oval or ellipsoidal central spore (most strains are spore forming) that does not distort the bacillary form.[1,22]

As said, most strains produce endospores, and this occurs within a few days of incubation on commonly used agar plates; such spores may be centrally or paracentrally placed, but they never distend the cell.[1] Through phase-contrast microscopy or techniques for spore staining, spore localization and morphology are much used criteria that may support discrimination among the species of the genus *Bacillus*.[1]

Finally, in wet preparations of clinical fluids or broth cultures, the peritrichous bacilli are found to be motile and to display a leisurely gait rather than a strong motility.[22]

*B. cereus* is usually referred to as a bacterium forming large, gray colonies (3–8 mm in diameter) with a rather flat and "ground-glass" morphology, frequently with irregular borders, and forming zones of β-hemolysis around colonies, when growing on agar media; the size of this hemolytic area is usually large, although it may vary based on growth conditions.[1] More exactly, when grown aerobically on 5% sheep blood agar at 37 °C, *B. cereus* colonies are dull gray colored; also, their aspect is opaque and a rough matted surface can be observed.[22] Colony edges are irregular and represent the configuration of bacterial swarming from the initial inoculum, perhaps due to swarming motility.[22] A β-hemolytic zone finally surrounds the colony, as mentioned above, although some strains are nonhemolytic.[23,24] Sometimes smooth colonies develop either alone or in the midst of rough ones. Notably, smears prepared from the frontal and distal advancing perimeters of a mature colony may reveal two diverse presentations. Smears from the distal edge show uniform rod-shaped forms with prominent central spores together with chains of Gram-positive rods, while those from the advancing edge are mostly made of masses of entangled chains of bacilli traversing the microscopic field; in this case, furthermore, absence of spore-containing bacilli may be appreciated. Perhaps, while spreading forward from the inoculum site, the colony leaves behind a trail of metabolic products that modify the pH and oxygen content in the growth medium, thus stimulating formation of spores. *B. cereus* also grows under an anaerobic atmosphere and at 45 °C, and biochemical identification may be achieved by combining the API 50 Carbohydrate (50 CH) and the API 20 Enterobacteriaceae (API 20 E) kits (both from bioMérieux, France).[1,25,51]

Phase-contrast microscopy and spore staining, which allow one to visualize spore localization and aspect, are used as criteria to discriminate among the species of the genus *Bacillus*, as said.[1] Other commonly used phenotype-based features that may be useful for identification are evaluation of motility, observation of hemolysis, and studying carbohydrate fermentation (*B. cereus* has been known not to ferment mannitol) and the very active lecithinase (phospholipase) activity.[1] Again, diverse and several plating media are available for *B. cereus* isolation, detection, and enumeration from food samples, including PEMBA (polymyxin–pyruvate–egg yolk–mannitol–bromothymol blue agar) and MYP (mannitol–egg yolk–phenol red–polymyxin agar).[1] Together with selective compounds like polymyxins, such media performance relies on *B. cereus* lecithinase production (egg-yolk reaction providing precipitate zones) and absence of fermentation of mannitol.[1] More recently, a number of chromogenic media have been designed for numerous food pathogens, including *B. cereus*, but conclusive identification should be based on molecular assays.[25] Methods for typing are discussed, however, in a separate chapter of this volume.

As *B. cereus* may be responsible for gas gangrene-like infections similar to those caused by the morphologically similar bacterium *Clostridium perfringens*, it is more than once required in clinical practice to promptly distinguish between them. In such cases, an India ink preparation can show encapsulated bacilli consistent with *C. perfringens*. Otherwise, lack of capsules is suggestive of *B. cereus* or maybe a non-*C. perfringens* clostridial species, although antibiotic coverage for *B. cereus* is needed, for prudence, pending culture results. Interestingly, the India ink test can be carried out directly with a tissue sample from the infected wound.[25]

## ANTIBIOTIC RESISTANCE

Usually, most *B. cereus* strains are resistant to penicillins and cephalosporins owing to a β-lactamase production. Given its potential seriousness, when a *B. cereus* infection is suspected prompt empirical treatment could be necessary pending antibiotic susceptibility testing (AST), and the choice must take into account such a common resistance trait involving β-lactam drugs. Furthermore, lack of susceptibility to macrolides, tetracyclines, and carbapenems has been described,[26,27] and the choice of a proper treatment is finally complicated by the fact that the European Committee for Antibiotic Susceptibility Testing (EUCAST), thus far, at the time of writing, has not yet published any guidelines for performing and interpreting AST with *Bacillus* spp. To address this issue, many investigators have undertaken *in vitro* susceptibility studies by using diverse methodologies aiming at providing some recommendations for an empirical antimicrobial approach. Nowadays, so, it seems that the drug of choice for *B. cereus* infections is vancomycin, while resistance to penicillin, ampicillin, cephalosporins (including broad-spectrum compounds), and trimethoprim is constant. Ciprofloxacin has, however, been shown to be greatly effective in the event of *B. cereus* wound infections,[28] and daptomycin and linezolid are reported to exert an almost uniform activity against this organism.

## CONCLUDING CONSIDERATIONS

*B. cereus* has evolved a panoply of pathogenic attributes, including adhesins and toxins, that make it able to enter and survive within the human host and, under certain circumstances, to breach barriers and cause pathologic processes in various anatomical compartments. These are discussed in a dedicated chapter, but it is important, here, to emphasize that the major hurdle, when *B. cereus* is isolated in clinical materials, is the need to overcome its nagging stigmata as a mere harmless contaminant, which is unfortunately still in vogue despite a globally ongoing evidence of diseases other than those (probably the most famous ones, together with ocular infections) involving the enteric tract.

Actually, outside the notoriety of *B. cereus* as an agent of food poisoning, acknowledgment of the multitude of other severe infections it may cause, such as fulminant bacteremias and devastating central nervous system involvement, is lacking. Both clinicians and clinical microbiologists must therefore seriously consider *B. cereus* isolates from clinical specimens before aprioristically dismissing them as accidentally contaminating organisms, especially when an underlying immunosuppression is present.

*B. cereus*, finally, by virtue of its exotoxin armamentarium, may be considered to be centrally situated between *B. anthracis* and, though an anaerobe, *C. perfringens*, thus representing a formidable link between these two sporulating bacterial species. In fact, *B. cereus* can both acquire and carry *B. anthracis* genes that are responsible for an anthrax-like pneumonia and, through its exoenzyme profile, overlap with *C. perfringens* pathogenicity in the event of gas gangrene-like skin syndromes.[25,29–32]

The debate surrounding classification of *B. cereus* group strains is not exclusively of academic and taxonomic interest, therefore; conversely, it does relate to issues concerning public health. In fact, for instance, *B. cereus* isolates that carry *B. anthracis* virulence determinants have caused cases of serious anthrax-like disease (see the dedicated chapter on *B. cereus* airway infections).[33] Again, while *B. cereus* is a well-known agent of food poisoning, *B. thuringiensis* plays a role as a biological insecticide for crop protection, as mentioned above.[1] Nevertheless, genes encoding the cytotoxins causing the diarrheal syndrome and other *B. cereus* infections are usually chromosomally encoded, and are present in all species of the *B. cereus* group, though silent in *B. anthracis*.[34] Especially, *B. thuringiensis* and *B. cereus* share similar distributions and expression of genes encoding extracellular pathogenic factors, and the former has caused human infections similar to those caused by the second.[35–41] Unfortunately, food poisoning caused by *B. thuringiensis* is probably underreported, given that methods for identification of *B. cereus* group strains cannot discriminate between *B. cereus* and *B. thuringiensis*.[42]

It has been suggested that *B. cereus*, *B. thuringiensis*, and *B. anthracis* be considered one species only, based on genetic evidence,[43] but there is no consensus on this matter thus far. Such an ambiguous taxonomy of the *B. cereus* group then reflects the difficulties encountered with species classification within bacterial systematics, especially today, in the so-called genomic era. It is also clear that phylogenetic analyses within the *B. cereus* group are furthermore complicated by extensive horizontal transfer of mobile genetic elements between strains.[44] Nevertheless, whereas the genome investigation suggests that the *B. cereus* group be considered as one species, a good argument supporting current nomenclature is the principle that "medical organisms with defined clinical symptoms may continue to bear names that may not necessarily agree with their genomic relatedness so as to avoid unnecessary confusion among microbiologists and nonmicrobiologists,"[45] as stated by rule 56a(5) in the *Bacteriological Code*.[46]

About the *B. cereus* phenotype, it is interesting to note, finally, that intermediate types between *B. cereus* and *B. weihenstephanensis* may sometimes be observed that are interpreted as a "snapshot" of ongoing thermal adaptation inside the *B. cereus* group.[47,48]

Phenotype may in fact not always reflect genotype; accordingly, and historically, *B. cereus*, *B. thuringiensis*, *B. anthracis*, and *B. mycoides* were described based on phenotype about one century ago, much earlier than the discovery of DNA as carrying heritable information. At that time, particularly, classification of bacteria was usually made on the basis of their morphological and physiological features, habitats, and pathogenicity for mammals or insects. Later, the large extent of synteny among their genomes and other genotypical similarities led scientists to suggest that the mentioned four species indeed represented a single taxon.[2,49]

In addition to the importance of *Bacillus* species as human pathogens as well as for agricultural purposes (related to their biocontrol properties, which are discussed in a separate chapter), numerous strains that have important economic relevance are used as probiotics, preventing enteric disorders.[4] Among such strains, *B. cereus* var. *toyoi* (strain BCT-7112T) has been used since 1975; in that year, in fact, it was officially approved by the Japanese Ministry of Agriculture and Forestry as the commercial product TOYOCERIN®. Spores of strain BCT-7112T have found application in animal nutrition, particularly for poultry, swine, cattle, and rabbits, along with aquaculture for over 30 years, in several countries throughout the world.[4] In the European Community, TOYO-CERIN® was first authorized by the European Commission in 1994 for use in swine, thus becoming the first bacterium authorized as a feed additive in the European Union; later, it was authorized for poultry, cattle, and rabbits, too.[4] Original identification studies assigned strain BCT-7112T to the species *B. cereus*; however, several phenotypic features can indeed differentiate it from the main phenotype of this species, so it was labeled as a variant and named *B. cereus* var. *toyoi*.[4] Following investigations based on DNA–DNA hybridization have documented that this organism might not be *B. cereus* after all, but actually represents a different species of the group, that is, *B. toyonensis*.[4]

## REFERENCES

1. Stenfors Arnesen LP, Fagerlund A, Granum PE. From soil to gut: *Bacillus cereus* and its food poisoning toxins. *FEMS Microbiol Rev* 2008;**32**:579–606.
2. Ash C, Farrow JA, Dorsch M, Stackebrandt E, Collins MD. Comparative analysis of *Bacillus anthracis, Bacillus cereus*, and related species on the basis of reverse transcriptase of 16S rRNA. *Int J Syst Bacteriol* 1991;**41**:343–6.
3. Savini V, Polilli E, Marrollo R, Astolfi D, Fazii P, D'Antonio D. About the *Bacillus cereus* group. *Intern Med* 2013;**52**:649.
4. Jiménez G, Urdiain M, Cifuentes A, López-López A, Blanch AR, Tamames J, et al. Description of *Bacillus toyonensis* sp. nov., a novel species of the *Bacillus cereus* group, and pairwise genome comparisons of the species of the group by means of ANI calculations. *Syst Appl Microbiol* 2013;**36**:383–91.
5. Frankland GC, Frankland PF. Studies on some new micro-organisms obtained from air. *R Soc Lond Phil Trans B* 1887;**178**:257–87.
6. Mock M, Fouet A. Anthrax. *Annu Rev Microbiol* 2001;**55**:647–71.
7. Jernigan DB, Raghunathan PL, Bell BP, Brechner R, Bresnitz EA, Butler JC, et al. Investigation of bioterrorism-related anthrax, United States, 2001: epidemiologic findings. *Emerg Infect Dis* 2002;**8**:1019–28.
8. Flugge C. *Die Mikroorganismen*. Leipzig (Germany): Vogel; 1886.
9. Nakamura LK. *Bacillus pseudomycoides* sp. nov. *Int J Syst Bacteriol* 1998;**48**:1031–5.
10. Johnson KM. *Bacillus cereus* food-borne illness. An update. *J Food Prot* 1984;**47**:145–53.
11. Sneath PHAClaus D, Berkeley RCW. Genus *Bacillus*. In: Sneath PHA, editor. *Bergey's manual of systematic bacteriology*, vol. 2. Baltimore (MD): Williams, Wilkins; 1986. p. 1105–39.
12. Lechner S, Mayr R, Francis KP, Pruss BM, Kaplan T, Wiessner-Gunkel E, et al. *Bacillus weihenstephanensis* sp. nov. is a new psychrotolerant species of the *Bacillus cereus* group. *Int J Syst Bacteriol* 1998;**48**:1373–82.
13. Kotiranta A, Haapasalo M, Kari K, Kerosuo E, Olsen I, Sorsa T, et al. Surface structure, hydrophobicity, phagocytosis, and adherence to matrix proteins of *Bacillus cereus* cells with and without the crystalline surface protein layer. *Infect Immun* 1998;**66**:4895–902.
14. Kotiranta A, Lounatmaa K, Haapasalo M. Epidemiology and pathogenesis of *Bacillus cereus* infections. *Microbes Infect* 2000;**2**:189–98.
15. Berkeley RCW, Logan NA, Shute LA, Capey AG. Identification of *Bacillus species*. In: Bergen T, editor. *Methods in microbiology*, vol. 16. London (United Kingdom): Academic Press; 1984. p. 291–328.
16. Anderson A, Granum PE, Rönner U. The adhesion of *Bacillus cereus* spores to epithelial cells might be an additional virulence mechanism. *Int J Food Microbiol* 1998;**39**:93–9.
17. Rönner U, Husmark U, Henrikson A. Adhesion of *Bacillus* spores in relation to hydrophobicity. *J Appl Bacteriol* 1990;**69**:550–6.
18. Jensen GB, Hansen BM, Ellenberg J, Mahillon J. The hidden lifestyles of *Bacillus cereus* and relatives. *Environ Microbiol* 2003;**5**:631–40.
19. Ghosh AC. Prevalence of *Bacillus cereus* in the faeces of healthy adults. *J Hyg (Lond)* 1978;**80**:233–6.

20. Vilain S, Luo Y, Hildreth MB, Brözel VS. Analysis of the life cycle of the soil saprophyte *Bacillus cereus* in liquid soil extract and in soil. *Appl Environ Microbiol* 2006;**72**:4970–7.
21. Turnbull PCB, Kramer JM. Intestinal carriage of *Bacillus cereus*: fecal isolation studies in three population groups. *J Hyg (Lond)* 1985;**95**:629–38.
22. Senesi S, Celandroni F, Scher S, Wong ACL, Ghelardi E. Swarming motility in *Bacillus cereus* and characterization of a *fliY* mutant impaired in swarm cell differentiation. *Microbiology* 2002;**148**:1785–94.
23. Turnbull PCB, Kramer J, Melling J. In: 8th ed. Topley WWC, Wilson GS, editors. *Topley and Wilson's principles of bacteriology, virology and immunity*, vol. 2. London (United Kingdom): Edward Arnold; 1990. p. 188–210.
24. Slamti L, Perchat S, Gominet M, Vilas-Boas G, Fouet A, et al. Distinct mutations in PlcR explain why some strains of the *Bacillus cereus* group are nonhemolytic. *J Bacteriol* 2004;**186**:3531–8.
25. Bottone EJ. *Bacillus cereus*, a volatile human pathogen. *Clin Microbiol Rev* 2010;**23**:382–98.
26. Kiyomizu K, Yagi T, Yoshida H, Minami R, Tanimura A, Karasuno T, et al. Fulminant septicemia of *Bacillus cereus* resistant to carbapenem in a patient with biphenotypic acute leukemia. *J Infect Chemother* 2008;**14**:361–7.
27. Savini V, Favaro M, Fontana C, Catavitello C, Balbinot A, Talia M, et al. *Bacillus cereus* heteroresistant to carbapenems in a cancer patient. *J Hosp Infect* 2009;**71**:288–90.
28. Turnbull PCB, Sirianni NM, LeBron CL, Samaan MN, Sutton FN, Reyes AE, et al. MICs of selected antibiotics for *Bacillus anthracis, Bacillus cereus, Bacillus thuringiensis*, and *Bacillus mycoides* from a range of clinical and environmental sources as determined by Etest. *J Clin Microbiol* 2004;**42**:3626–34.
29. Jernigan JA, Stephens DS, Ashford DA, Omenaca C, Topiel MS, et al. Bioterrorism-related inhalational anthrax: the first 10 cases reported in the United States. *Emerg Infect Dis* 2001;**7**:933–44.
30. Walker D. Sverdlovsk revisited: pulmonary pathology of inhalational anthrax versus anthrax-like *Bacillus cereus* pneumonia. *Arch Pathol Lab Med* 2012;**136**:235.
31. Wright AM, Beres SB, Consamus EN, Long SW, Flores AR, Barrios R, et al. Rapidly progressive, fatal, inhalation anthrax-like infection in a human: case report, pathogen genome sequencing, pathology, and coordinated response. *Arch Pathol Lab Med* 2011;**135**:1447–59.
32. Hoffmaster AR, Ravel J, Rasko DA, et al. Identification of anthrax toxin genes in a *Bacillus cereus* associated with an illness resembling inhalation anthrax. *Proc Natl Acad Sci USA* 2004;**101**:8449–54.
33. Mignot T, Mock M, Robichon D, Landier A, Lereclus D, Fouet A. The incompatibility between the PlcR- and AtxA-controlled regulons may have selected a nonsense mutation in *Bacillus anthracis*. *Mol Microbiol* 2001;**42**:1189–98.
34. Damgaard PH. Diarrhoeal enterotoxin production by strains of *Bacillus thuringiensis* isolated from commercial *Bacillus thuringiensis*-based insecticides. *FEMS Immunol Med Microbiol* 1995;**12**:245–50.
35. Rivera AMG, Granum PE, Priest FG. Common occurrence of enterotoxin genes and enterotoxicity in *Bacillus thuringiensis*. *FEMS Microbiol Lett* 2000;**190**:151–5.
36. Swiecicka I, Van der Auwera GA, Mahillon J. Hemolytic and nonhemolytic enterotoxin genes are broadly distributed among *Bacillus thuringiensis* isolated from wild mammals. *Microb Ecol* 2006;**52**:544–51.
37. Samples JR, Buettner H. Corneal ulcer caused by a biologic insecticide (*Bacillus thuringiensis*). *Am J Ophthalmol* 1983;**95**:258–60.
38. Jackson SG, Goodbrand RB, Ahmed R, Kasatiya S. *Bacillus cereus* and *Bacillus thuringiensis* isolated in a gastroenteritis outbreak investigation. *Lett Appl Microbiol* 1995;**21**:103–5.
39. Hernandez E, Ramisse F, Ducoureau JP, Cruel T, Cavallo JD. *Bacillus thuringiensis* subsp. *konkukian* (serotype H34) superinfection: case report and experimental evidence of pathogenicity in immunosuppressed mice. *J Clin Microbiol* 1998;**36**:2138–9.

40. Ghelardi E, Celandroni F, Salvetti S, Fiscarelli E, Senesi S. *Bacillus thuringiensis* pulmonary infection: critical role for bacterial membrane-damaging toxins and host neutrophils. *Microbes Infect* 2007;**9**:591–8.

41. Granum PE. *Bacillus cereus* and food poisoning. In: Berkeley R, Heyndrickx M, Logan N, De Vos P, editors. *Applications and systematics of Bacillus and relatives.* Oxford: Blackwell Science Ltd; 2002. p. 37–46.

42. Helgason E, Økstad OA, Caugant DA, Johansen HA, Fouet A, Mock M, et al. *Bacillus anthracis, Bacillus cereus,* and *Bacillus thuringiensis* – one species on the basis of genetic evidence. *Appl Environ Microbiol* 2000;**66**:2627–30.

43. Cardazzo B, Negrisolo E, Carraro L, Alberghini L, Patarnello T, Giaccone V. Multiple-locus sequence typing and analysis of toxin genes of *Bacillus cereus* foodborne isolates. *Appl Environ Microbiol* 2008;**74**:850–60.

44. Stackebrandt E, Frederiksen W, Garrity GM, et al. Report of the ad hoc committee for the re-evaluation of the species definition in bacteriology. *Int J Syst Evol Microbiol* 2002;**52**:1043–7.

45. Lapage SP, Sneath PHA, Lessel EF, Skerman VBD, Seeliger HPR, Clark WA. *International code of nomenclature of bacteria (1990 revision). Bacteriological code.* Washington, DC: American Society for Microbiology; 1992.

46. von Stetten F, Mayr R, Scherer S. Climatic influence on mesophilic *Bacillus cereus* and psychrotolerant *Bacillus weihenstephanensis* populations in tropical, temperate and alpine soil. *Environ Microbiol* 1999;**1**:503–15.

47. Stenfors LP, Granum PE. Psychrotolerant species from the *Bacillus cereus* group are not necessarily *Bacillus weihenstephanensis*. *FEMS Microbiol Lett* 2001;**197**:223–8.

48. Stenfors LP, Mayr R, Scherer S, Granum PE. Pathogenic potential of fifty *Bacillus weihenstephanensis* strains. *FEMS Microbiol Lett* 2002;**215**:47–51.

49. Rasko DA, Altherr MR, Han CS, Ravel J. Genomics of the *Bacillus cereus* group of organisms. *FEMS Microbiol Rev* 2005;**29**:303–29.

50. Aronson AI, Shai Y. Why *Bacillus thuringiensis* insecticidal toxins are so effective: unique features of their mode of action. *FEMS Microbiol Lett* 2001;**195**:1–8.

51. Logan NA, Berkeley RCW. Identification of *Bacillus* strains using the API system. *J Gen Microbiol* 1984;**130**:1871–82.

CHAPTER 2

# *Bacillus cereus* Group Diagnostics: Chromogenic Media and Molecular Tools

## Eugenio Pontieri

Department of Biotechnological and Applied Clinical Sciences, University of L'Aquila, Italy

## INTRODUCTION

In old publications, attention focused on the genus *Bacillus* gave a good idea of the differences between the strains and species belonging to this genus.[33,76] *Bacillus cereus* was classified as a large-celled species of Group 1 (with a cell width greater than 0.9 mm and whose spores do not appreciably swell the sporangium). The genus *Bacillus* consists of Gram-positive, spore-forming, highly heterogeneous bacteria, often strictly connected to animal or human pathogenicity. *Bacillus cereus* is one of the most important members of the genus, notably as a potential food pathogen; *Bacillus anthracis* was first identified by Koch in 1876 as a pathogenic bacterium in animals and humans responsible for anthrax[47]; and *Bacillus thuringiensis* is considered a biopesticide. The high genetic relatedness among these three principal species has contributed to the idea that they are a single species, *B. cereus sensu lato* (*s.l.*).[12,15,38,48] Differences among *B. cereus*, *B. anthracis*, and *B. thuringiensis* have been found by specific phenotypic characteristics and, in particular, pathogenic properties.

*Bacillus anthracis* is the etiologic cause of anthrax, affecting animals (primarily herbivores) and humans, the spores of which could be used in bioterrorism.[41,42] The bacterium fully virulent possesses two large plasmids, pX01 and pX02, codifying respectively the subunits of the whole anthrax toxin and the poly-$\gamma$-D-glutamic capsule (uniquely proteinaceous among bacteria).

*Bacillus cereus sensu stricto* (*s.s.*) possesses enterotoxins able to confer the emetic or diarrheic form of gastroenteritis.[6] These toxins have also been detected in *Bacillus* spp. not belonging to the *B. cereus* group.[60,66] A heat-stable toxin related to the *B. cereus* emetic toxin cereulide has been found in *Paenibacillus tundrae* by Rasimus et al.[61] Strains or species belonging to the

*The Diverse Faces of Bacillus cereus*
ISBN 978-0-12-801474-5
http://dx.doi.org/10.1016/B978-0-12-801474-5.00002-5

*Bacillus* genus can produce several enzymes contributing to food spoilage, which may vary among strains of the same species.[16] Human infections have also been reported.[69,88]

*Bacillus thuringiensis*, considered an insect pathogen,[63] has been utilized commercially as an insect biopesticide, in particular for *Lepidoptera*, *Diptera*, and *Coleoptera*,[7,55] and is also used in control of the vectors for malaria and yellow fever.[57] The spores of this bacterium are associated with inclusions by a large crystal protein of 130–140 kDa as a protoxin, becoming a 60-kDa δ-endotoxin after ingestion in the gut of insects. The *cry* genes are located on large transmissible plasmids. The formation of protein crystals during sporulation can distinguish *B. thuringiensis* from *B. cereus*. A *B. cereus* bacterium that is acrystalliferous but possessing the *cry* genes is considered a *B. thuringiensis*. A case of pulmonary infection by *B. thuringiensis* has also been reported.[32]

Other members of the group are *Bacillus mycoides*, *Bacillus pseudomycoides*, and *Bacillus weihenstephanensis*, which seem to be of low pathogenic potential, but are psychrotolerant, being able to grow under refrigerator conditions, and of spoilage potential.[35,36] *Bacillus mycoides* is recognized by the rhizoid growth resembling a fungus on agar plates. Another member, *Bacillus cytotoxicus*, has been recognized as a potential food-poisoning pathogen.[37]

With the development of new and finer molecular techniques, especially whole genome sequencing, new descriptions of "borderline" strains have been reported, making it necessary to reconsider the nomenclature.[29,45]

Considering the pathogenic potential of the entire *B. cereus* group, it is necessary to obtain a better understanding of the genomes belonging to the group, by whole sequencing and gene analysis. An overview of classical and molecular analysis, such as culture and chromogenic media and molecular tools, will be provided in this chapter, focusing on their diversity, sensitivity, and capacity to reveal toxic and nontoxic strains.

## CULTURE AND CHROMOGENIC MEDIA

If not necessary, special handling processing is not required, but, obviously, major cautions are due *B. anthracis* and in minor entity to *B. cereus*. A Gram staining can be applied to specimens for a first view. This step can differentiate among the various species with respect to aspect and disposition of cells and spores (Table 1). As reported in manuals of diagnostic microbiology (see Ref. 80), the first medium to analyze the isolates may be 5% sheep blood agar plates at 37 °C or other blood culture medium. In these plates, we can observe the first differences among the *Bacillus* spp. because *B. anthracis* is

**Table 1** Characteristics of *Bacillus cereus* group species

| Feature | Bacillus cereus | Bacillus cereus var. mycoides | Bacillus thuringiensis | Bacillus anthracis |
|---|---|---|---|---|
| Gram stain | + | + | + | + |
| Catalase | + | + | + | + |
| Egg yolk reaction | ± | ± | ± | (+) |
| Motility | ± | − | ± | − |
| Acid from mannitol | − | − | − | − |
| Hemolysis (sheep red blood cells) | + | (+) | + | − |
| Rhizoid growth | − | + | − | − |
| Toxin crystals produced | − | − | + | − |
| Glucose anaerobic utilization | + | + | + | + |
| Reduction of nitrate | ± | + | + | + |
| VP[a] reaction | + | + | + | + |
| Tyrosine decomposition | + | (+) | + | (+) |
| Resistance to lysozyme | + | + | + | + |

[a]Voges-Proskaurer test.
+, positive; ±, usually positive but occasionally may be negative; (+), often weakly positive; −, negative.
From Vanderzant C, Splittstoesser DF, editors. *Compendium of methods for the microbiological examination of foods*. 3rd ed. American Public Health Association; 1992.

nonhemolytic (Table 1), owing to a nonsense mutation in the pleiotropic regulator gene (*plcR*).[2,75]

In his review, E.J. Bottone[6] reported the appearance at 37 °C of *B. cereus* colonies on 5% blood sheep agar. They are dull gray and opaque with a rough matted surface. Irregularity of colony perimeters is visible, representing the configuration of swarming from the site of initial inoculation, perhaps due to *B. cereus* swarming motility.[71] Zones of β-hemolysis are visible around colonies and conform to the colony morphology.[87] In some instances, smooth colonies develop, either alone or in the midst of rough colonies. Grown apart from the initial inoculum, smooth colonies are surrounded by a uniform zone of β-hemolysis framing the centrally situated colony. Furthermore, smears prepared from the distal and frontal (spreading) advancing perimeters of a mature colony may reveal two distinct morphological presentations. Smears prepared from the distal edge show uniform bacillary forms with prominent centrally situated spores admixed with chains of Gram-positive bacilli, while smears from the advancing edge comprise predominately masses of entangled bacillary chains traversing the microscopic field and a remarkable absence of spore-containing bacilli. Perhaps, as the colony spreads forward from the initial inoculum site, it leaves behind a trail of metabolic end products, which alters the pH and oxygen content of the growth environment, thereby inducing spore formation. *B. cereus* grows anaerobically and at 45 °C. After the direct examination of the colony aspect on blood agar plates a biochemical characterization of the isolates of *Bacillus* spp. can be obtained through the use of API 20 Enterobacteriaceae (API 20 E) and API 50 Carbohydrate (50 CH) kits (bioMérieux, France), used together[53] (Tables 1 and 2).

Other general media can be utilized, such as Columbia agar supplemented with nalidixic acid and colistin (attention is due because some isolates can be sensitive to nalidixic acid), phenylethyl alcohol agar and polymyxin–lysozyme–EDTA–thallous acetate, all media selective for Gram-positive microorganisms. The colonies on these types of plates appear creamy white, circular, and domed. In addition, we can utilize media to induce the proteinaceous capsule formation in *B. anthracis*, such as bicarbonate agar.[80]

Apart from *B. anthracis*, the most important cultural techniques for isolation and enumeration of *B. cereus* group species are ruled by International Organization for Standardization (ISO) standards ISO 7932 and ISO 21871[4] and the Food and Drug Administration (FDA).[31] Mannitol–egg yolk–polymyxin (MYP; Oxoid) and polymyxin–egg yolk–mannitol–bromothymol blue (PEMBA; Oxoid) agars are still used in the standard procedure for detection and enumeration of presumptive *B. cereus s.l.* in food matrix. In MYP and PEMBA, typical colonies of the *B. cereus* group strains are recognized by a precipitation zone due to egg yolk hydrolysis and, respectively, appear colored by pink or peacock blue. However, in these media the risk of misidentification of various strains, in particular from food matrices, is possible; atypical reactions on these media are reported by Fricker et al.[29]

To better identify *B. cereus* group bacteria, new chromogenic media have been developed, such as the Chromogenic *Bacillus cereus* agar (CBC; Oxoid). This medium contains 5-bromo-4-chloro-3-indolyl-β-D-glucopyranoside cleaved by the β-D-glucosidase. The resulting colonies are white with a blue-green center. BCM® (Biosynth AG, Switzerland) is another chromogenic medium. The chromogenic substrate is 5-bromo-4-chloro-3-indoxyl myoinositol-1-phosphate, which, if cleaved by phosphatidylinositol-phosphatase, confers a blue-turquoise color to *B. cereus* colonies, sometimes surrounded by a blue halo.[59] Biolife (Milan, Italy) produces another chromogenic medium. In this medium the *B. cereus* group isolates are visible as pink-orange colonies surrounded by a halo of hydrolyzed egg yolk.

Fricker and coworkers in 2008 compared PEMBA, MYP, CBC, and BCM, testing a strain set of 100 *B. cereus* with different origins (food isolates,

**Table 2** *Bacillus* species differentiation detected in clinical specimens[a]

| Species | β-Hemolysis | Motility | Lecithinase | Gelatin hydrolysis | Anaerobic growth | Nitrate reduction |
|---|---|---|---|---|---|---|
| *Bacillus anthracis* | 0 | 0 | + | V | + | + |
| *Bacillus cereus* | + | + | + | + | + | V |
| *Bacillus licheniformis* | + | + | 0 | + | + | + |
| *Bacillus megaterium* | 0 | + | 0 | + | 0 | 0 |
| *Bacillus mycoides* | 0 | 0 | + | + | + | V |
| *Bacillus pumilus* | V | + | 0 | + | 0 | 0 |
| *Bacillus sphaericus* | 0 | + | 0 | + | V | |
| *Bacillus subtilis* | V | + | 0 | + | 0 | + |
| *Bacillus thuringiensis* | + | + | + | + | + | + |

[a]+, positive reaction (≥85%); V, variable reaction (positive 15–84%); 0, negative reaction (positive <15%).

food-borne outbreaks, and clinical isolates) and toxigenic potentials. In addition, naturally contaminated foods were utilized to test the performance of the four plating media in analysis of complex samples. They found that conventional media could lead to a substantial misidentification and underestimation of food illness by *B. cereus*, especially when used by laboratory staff not highly trained in this identification. Moreover, some strains may not be detected, even in the two chromogenic media, in particular a highly toxic strain. They concluded that different techniques are necessary for correct identification. In particular, sequence variances in the *plcR* gene (the pleiotropic regulator of various virulence factors) are strictly linked with atypical colony aspects.

For the isolation and enumeration of presumptive *B. cereus* groups in food, two methods can be utilized, a colony-count technique on solid agar of various media or the most probable number technique. Normally this is the procedure: On the surface of two agar plates, 0.1 ml of the sample is distributed in duplicate, using a liquid or a stock suspension, for nonliquid products, followed by successive dilutions up to $10^{-6}$. To increase the sensitivity 1 ml of sample may be plated on a 150-mm petri dish. Incubate at $30 \pm 1\,°C$ aerobically for $24 \pm 2\,h$. Count the typical colonies of *B. cereus* in plates containing 15 to 150 colonies, considering as such the growths that have the expected characteristics. The minimal infectious dose is considered to be $100\,\text{cell}\,g^{-1}$ but infection can occur to a number of 10,000 cells.

Another technique to distinguish *B. anthracis* from other *Bacillus* spp., in particular *B. cereus*, is to test the isolates with a suspension of the gamma phage.[1] This technique is considered of definitive determination also against *B. anthracis*-like phenotype white colonies in BCM plates,[29] but in the same work a *B. weihenstephanensis*-type strain showed variable results in a gamma phage assay.

## MOLECULAR DIAGNOSTIC TOOLS

Alongside the cultural and chromogenic typing tools, molecular typing approaches have been developed for better discrimination among *Bacillus* species and/or strains or to conduct epidemiological studies in outbreak situations and to monitor contamination routes.

### Species and Strain Identification

Classical cultural detection, enumeration, and differentiation cannot discriminate highly among isolates of *B. cereus* groups, so combining them with molecular-based tools can enhance this capacity. The pathogenic properties of the *B. cereus* group strains are due to specific toxin genes they possess. These genes can be detected with standard molecular techniques such as DNA amplification and sequencing.

Real-time polymerase chain reaction (PCR) assays have been applied in the enumeration of *B. cereus* strains,[10,18,54] but difficulties arise because of the presence of several states of the cells, vegetative or sporal, live or dead.

PCR identification methods have been developed especially for *B. anthracis* because of its bioterrorism potential. Wielinga et al.[91] developed a multiplex real-time PCR for identifying and differentiating *B. anthracis* virulence types. They integrated three markers: (1) the coding region of the edema factor gene (*cya*; component of the anthrax toxin), placed on pXO1; (2) the coding region of the capsule synthesis gene *capB*, located on pXO2; and (3) an internal chromosomal marker sequence. Previously Leski et al.[51] utilized the *Bacillus* collage-like protein *bcl* gene as the target sequence in the detection and fingerprinting of *B. anthracis* with respect to the other *B. cereus* group species.

PCR detection protocols for profiling the toxin genes of the *B. cereus* group have been developed utilizing classic gel-based PCR or real-time PCR.[19,22,28,34,90] Owing to the possession of more than one toxin by members of the *B. cereus* group, diagnostic tools for toxin profiling should include genes encoding the three major toxins, the nonhemolytic entero-toxin (Nhe), the hemolysin BL (Hbl) and the cytotoxin K (CytK) adding the emetic toxin cereulide synthase gene *ces* that has been found har-boured on a large plasmid.[20,22,77] The *nhe* genes are carried by all the *B. cereus* group strains, but not all strains produce Nhe, and levels can vary in a temperature-dependent manner as found by Réjasse et al.[64] in the psy-chrotolerant *B. weihenstephanensis*, owing to the expression of the *plcR* gene. *Bacillus cereus* strains produce the toxin genes at various levels.[24] Especially in the food industry and food microbiology there is the possi-bility of the emergence of new phenotypes carrying novel toxin gene profiles, as reported by several authors.[13,23,81]

Toxin gene profiling is an analysis much more important with respect to species determination, especially in outbreak situations. In fact, not only does *B. cereus s.s.* harbor the previously mentioned toxin genes, but they can also be distributed among the *B. cereus* group.[35,40,81] In summary, in food microbiology the diagnostics should be addressed more to the determina-tion of the toxin or virulence genes; the differentiation of the species does not seems to be of primary importance. Moreover, a toxin quantification method should be added to the toxin gene determination.[5]

## Epidemiological Studies

In epidemiological studies, for monitoring an outbreak of the food pathogen *B. cereus*, a fingerprinting technique can be requested. Pulsed-field gel electrophoresis (PFGE) is considered one of the most important fingerprint typing methods in microbiologic analysis of pathogens isolated from any type of source (clinical, environmental, food). *Bacillus cereus* was first analyzed by PFGE.[9,52,58] From analysis of undigested DNA blocks by PFGE, all *B. cereus* strains showed one or more extrachromosomal bands migrating to a region corresponding to 6 kb or larger,[9] variable among the isolates, probably corresponding to plasmids. DNA blocks normally are digested with specific restriction enzymes, resulting in a macrorestriction endonuclease fingerprinting able to discriminate among the isolates. Actually, *Sma*I and *Apa*I enzymes are considered the best choice.[52,58] To better analyze food pathogens in particular, the International Molecular Subtyping Network for Foodborne Disease Surveillance, known simply as PulseNet International,[79] has been created. As written on the site's Home page, PulseNet International is a network of national and regional laboratory networks dedicated to tracking food-borne infections worldwide. Each laboratory utilizes well-standardized genotyping methods, sharing information in real time. The resulting surveillance provides early warning of food- and waterborne disease outbreaks, emerging pathogens, and acts of bioterrorism. Several well-known food pathogens, especially Gram-negative ones, are the object of the site, including *Listeria monocytogenes*. No specific protocols for *Bacillus* spp. are present, but protocols described for *L. monocytogenes* may be usefully applied.

## Population Studies and Contamination Route Analysis

PCR-based methods are available for molecular typing of *B. cereus* groups. The randomly amplified polymorphic DNA (RAPD) assay represents a valid and largely used tool for molecular typing of various *Bacillus* species and can be utilized routinely in the laboratory as a screening method. In particular, RAPD has been utilized in typing *B. anthracis* with respect to other members of the *B. cereus* group and for epidemiological study of *B. cereus* and *Bacillus lentus*.[14,78] Other techniques are utilized in the screening of repetitive elements, such as repetitive extragenic palindromic, enterobacterial repetitive intergenic consensus, and BOX element PCR.[24]

In particular, the PulseNet Web site suggests the multiple-locus variable-number tandem repeat analysis (MLVA). Short repeat sequences are dispersed throughout the genome in virtually all prokaryotic and eukaryotic genomes. In each repeat sequence locus the repeat copy number can vary between strains; hence these sequences are often referred to as "variable number tandem repeats" or VNTRs. If multiple different VNTRs are targeted for analysis the technology is called "multiple-locus VNTR analysis" or MLVA. In MLVA, the VNTR array is amplified using PCR.[56] This technique is consistent with the great genetic variability found between the *B. anthracis* and the *B. cereus* strains.[11,44] The analysis of single-nucleotide polymorphisms (SNPs) is being evaluated at the Centers for Disease Control (CDC) as a subtyping platform to identify specific subtypes or characteristics of bacterial isolates. VNTRs and SNPs are considered targets of choice as described by Keim et al.[43] and Kuroda et al.[49] for typing isolates belonging to genetically homogeneous *B. anthracis* species.

In collaboration with the U.S. Department of Agriculture (USDA), PulseNet has also begun to evaluate multilocus genome typing for high-resolution subtyping of *Listeria* spp. The application of these technologies also to *Bacillus* species will complement existing PFGE-based methods, leading to faster, more accurate data acquisition and an increased likelihood of identifying an outbreak earlier so that effective responses may be implemented.

Multilocus sequence typing (MLST), however, developed in the early 1990s,[26] has gradually become the "gold standard" in molecular typing, becoming a central genotyping strategy for analysis of bacterial populations. The revealed genomic plasticity confers a dynamism to the population structure of *B. cereus s.l.*, qualifying the possibility of expressing new strains with incremented or new virulence or a capacity to better survive under adverse environmental conditions.[23,48,85] Applied to the genus *Bacillus* and in particular to *B. cereus s.l.*, MLST has focused the relationship among isolates from various isolation sources,[8,21,39,62,83,89] revealing the existence of three major clades. Normally, conserved oligonucleotide primers are designed to amplify 300- to 600-bp fragments of six to eight housekeeping genes. However, MLST has been limited because it is expensive and has several drawbacks (the purchase, maintenance, and housing of several pieces of laboratory equipment and tracking of data). The development of microfluidic biochips might simplify MLST analysis in the future.[65] Any typing method used has shown the various *B. cereus* group species distributed within three major clusters, making a diagnostic based solely on species identification insufficient. Interestingly, Ehling-Schulz and coworkers[21] found the same major clusters by Fourier-transform infrared spectroscopic analysis combined with RAPD profiles screening 90 isolates of different geographic origins. In particular, they that showed emetic toxin formation of B. *cereus* is restricted to a single evolutionary lineage of closely related strains. Moreover, the use of the sporulation stage III AB gene (*spoIIIAB*) as a single genetic marker, as in this study, might represent an alternative to obtain a rough snapshot of genetic relations among B. *cereus s.l.* strains.[21] This genetic marker resembles the structure of MLST-derived clusters and its suitability for sequence typing was reconfirmed by comparing clusters derived from hierarchical cluster analysis of *spoIIIAB* sequences, with the clusters obtained by whole genome sequencing using a sliding window approach.[30,70]

For some purposes, Amplified Fragment Length Polymorphism (AFLP) could be preferred to MLST, especially if high throughput capacities are needed, avoiding biases due to potentially inadequate MLST schemes. Guinebretière et al.[35] had associated the fluorescent AFLP (fAFLP), techniques based on genome-wide blind markers, to phylogeny of ribosomal genes and the housekeeping gene *panC* to the characterization of 425 strains of B. *cereus* Group from very different ecological niches. They identified seven major clusters (denoted I–VII) correlating with physiological properties of the strains. Interestingly, the potential of the strains for causing food poisoning correlated with certain phylogenetic groups.[36] The data collected in this type of analysis have to be processed by a specific software.

Online tools have been developed for attributing strains to different genetic groups: (1) https://www.tools.symprevius.org/Bcereus/english.php permits the affiliation to phylogenetic groups (I to VII) within the B. *cereus* group.[35] (2) HyperCat (http://mlstoslo.uio.no)[84] allows the integration of data from two different typing systems (MLST, AFLP) as well as data derived from multilocus enzyme electrophoresis, to calculate supertrees. HyperCat was applied to carry out multidata type analysis on 2213 strains of different origins, including 450 food and dairy production strains. This integrative approach confirmed the major clusters but also revealed some novel phylogenetic branches, including a putative new lineage of B. *anthracis*.[85] More data integration from functional genomics (transcriptomics, proteomics, and metabolomics) is the next step to obtain a more holistic understanding of this important group of microorganisms.

## Matrix-Assisted Laser Desorption Ionization Time-of-Flight Mass Spectrometry

Matrix-assisted laser desorption ionization time-of-flight mass spectrometry (MALDI-TOF-MS) is a technique that has been successfully adapted and implemented for the routine identification of microorganisms in clinical microbiology laboratories since 2005. This revolutionary technique allows for easier and faster diagnosis of human pathogens with respect to conventional phenotypic and molecular identification tools, yielding unquestionable reliability and cost-effectiveness[73,74]

Several authors have reported in several studies the identification of various species-specific biomarkers from *B. anthracis* spores and vegetative forms.[17,25,46,50,67,68] Ryzhov et al.[67] utilized MALDI-TOF-MS to characterize spores of 14 microorganisms of the *B. cereus* group inclusive of the four species *B. anthracis*, *B. cereus*, *B. mycoides*, and *B. thuringiensis*. MALDI mass spectra obtained from whole bacterial spores showed many similarities between the species, except for *B. mycoides*. At the same time, unique mass spectra were obtained for the different *B. cereus* and *B. thuringiensis* strains, allowing for differentiation at the strain level. Treating the spores with corona plasma discharge or sonication prior to MALDI analysis, they obtained an increase in the number of detectable biomarkers in the usually peak-poor MALDI spectra of spores, displaying an ensemble of biomarkers common for *B. cereus* group bacteria and useful for differentiating *B. cereus* group spores from those of *Bacillus subtilis* and *Bacillus globigii*. Three peculiar biomarkers were found by Elbanany et al.[25] using MALDI-TOF-MS able to differentiate spores of *B. anthracis* with respect to the three closely related species *B. cereus*, *B. thuringiensis*, and *B. mycoides*. Lasch and coworkers[50] reported the use of MALDI-TOF-MS to identify *B. anthracis*. Mass spectra of 102 *B. anthracis* isolates, 121 *B. cereus* isolates, and 151 other *Bacillus* and related genera isolates were analyzed. Mass spectra analysis by gel view and unsupervised hierarchical cluster analysis permitted the classification of *Bacillus* strains into two groups (*cereus* and non-*cereus*), and *B. anthracis* were correctly classified into two different clusters of six subgroups of the *B. cereus* group. Classification models using artificial neural networks created from 296 mass spectra were used to blindly identify 100% of *B. anthracis*, 94.5% of *B. cereus*, and 92.9% of *Bacillus* non-*cereus* strains. However, the complexity of these data analysis methods prevents their application in clinical laboratories for routine diagnoses. It is evident that the lack of suitable tools for studying enterotoxins hampers the possibilities for accurate hazard identification and characterization in microbial food-safety risk assessment.

More recently, MALDI-TOF–MS was applied to the detection of enterotoxins produced by pathogenic strains of the *B. cereus* group.[86] Enterotoxins produced by various species of the *B. cereus* group, such as CytK1 and Nhe, have been associated with diarrheal food poisoning incidents. CytK1 detection is not possible with commercial assays, whereas Nhe is recognized by the ELISA-based 3M Tecra *Bacillus* Diarrhoeal Enterotoxin Visual Immunoassay, which does not specifically target this protein because it is based on polyclonal antibodies.

The authors have applied MALDI-TOF–MS for the detection of CytK1 and Nhe produced by pathogenic strains of the *B. cereus* group using protein digests from one-dimensional gel electrophoresis. Secretion of CytK1 and two of the three components of Nhe was confirmed in supernatants of various *B. cereus* cultures. For each protein, biomarkers were introduced that could be used for the screening of food poisoning or food/environmental isolates able to secrete enterotoxins. Therefore, MALDI-TOF–MS could be successfully utilized as a risk assessment procedure to determine the involvement of strains of the *B. cereus* group in food-borne outbreaks, including the CytK1-producing species *B. cytotoxicus*, described in 2012.[86]

The cost of bacterial identification by MALDI-TOF-MS has been evaluated at €1.43 compared with €4.60–8.23 for conventional identification studies.[72] The assay time is decidedly decreased, 6–8.5 min for bacterial identification from a direct colony versus 5–48 h for conventional identification. New specific databases may improve the performance of MALDI-TOF-MS for routine bacterial identification. However, the cost of the apparatus, ranging from €100,000 up to €200,000, and the absence of specific reagents at this time limit the use of MALDI-TOF–MS in the routine microbiology laboratory.

## THE FUTURE OF DIAGNOSTICS

In the past years, massive efforts have been made in genome sequencing. New sequencing hardware and software data analysis have enabled the sequencing of whole microbial genomes in less than 1 day. In particular for the *B. cereus* group, the role of *B. anthracis* as a potent agent in bioterrorism and the finding of *B. cereus s.s.* toxins in other correlate strains has renewed interest in this group of bacteria leading to several genome sequencing projects. At this time (September 2014), genomic sequence information is available for about 327 *B. cereus* group strains and is increasing (Table 3). However, considering the high rate of genome plasticity for bacteria belonging to the *B. cereus* group,[23,48,82] de novo sequence assembly is laborious and time-consuming. For this reason, the appropriate development of easy-to-use bioinformatic tools is now a dynamic trend. The first tools minimizing post-sequencing data processing have already been developed. Segerman et al.[70] obtained a subtyping of *B. anthracis* strains and a general overview of the

**Table 3** *Bacillus cereus* group multiple genomes published on http://www.ncbi.nlm.njh.gov/genome [accessed September 2014]

| Organism | Size (Mb) | Chrs | Plasmids | Assemblies |
|---|---|---|---|---|
| *Bacillus cereus* group | 6.58303 | 1 | 1 | 67 |
| *Bacillus anthracis* | 5.9473 | 1 | 2 | 53 |
| *Bacillus cereus* | 6.85877 | 1 | 7 | 153 |
| *Bacillus cytotoxicus* | 4.09416 | 1 | 1 | 1 |
| *Bacillus gaemokensis* | 5.59986 | – | – | 1 |
| *Bacillus manliponensis* | 4.64311 | – | – | 1 |
| *Bacillus mycoides* | 5.8415 | 1 | 5 | 4 |
| *Bacillus thuringiensis* | 7.08969 | 1 | 28 | 41 |
| *Bacillus toyonensis* | 5.02542 | 1 | 2 | 1 |
| *Bacillus weihenstephanensis* | 5.90241 | 1 | 4 | 5 |

phylogenomic structure of the genus *Bacillus*. Instead of assembling the genome, defining genes, defining orthologous relations, and calculating distances, the method can achieve a similar high resolution directly from the raw sequence data. Agren et al.[3] presented a new software tool, "Gegenees," as a fragmented alignment of multiple genomes for determining phylogenomic distances and genetic signatures unique to specified target groups. The site is accessible as an open platform at http://www.gegenees.org/index.html. The software was successfully used to search for unique signatures for *B. anthracis* by analyzing 134 *Bacillus* genomes. Target group-specific primers were designed by using the identified signatures.[3]

DNA microarray-based analysis represents another method for following genomic typing. The method improves with acquisition of increasing data of *Bacillus* genomic sequencing projects, making it possible to design high-density whole genome microarrays amplifying the ability to discriminate among strains. For instance, Felder et al.[27] used, in a multigenome DNA array of virulence genes, the genomic marker gene *rpoB*, as well as the selective 16S rDNA sequence regions of *B. anthracis*, the *B. cereus* group, and *B. subtilis*. Eight *B. anthracis* reference strains were tested and correctly identified. Among the analyzed environmental *Bacillus* isolates, no virulent *B. anthracis* strain was detected. Moreover, any other type of gene increasing the pathogenic potential could be implemented in microarray analysis to discriminate better among *Bacillus* isolates or strains. Last, but not least, MALDI-TOF-MS represents a new exciting diagnostic tool, especially if new specific databases can be implemented in routine bacterial identification and the whole cost of the apparatus and utilization decreases.

## CONCLUSIONS

The *B. cereus* group comprises strains diversely involved in human pathogenesis, owing to the production of several highly toxic proteins, together with strains not pathogenic or even beneficial. The detection of strains belonging to this group has become a major aim in diagnostics microbiology. Species and/or strain discrimination among the *B. cereus* group strains may be not sufficient. Detection and quantification of toxins responsible for pathogenesis and characterization of markers for monitoring contamination routes appear more important. During the past years, to satisfy all these needs, impulse has been given to developing new molecular tools leading to a significant improvement in *B. cereus* group diagnostics. However, the classical cultural methods for detection and/or enumeration of members of the *B. cereus* group, improved by new chromogenic media, are still important tools in food microbiology as ruled in food analysis international protocols. The new molecular tools provide a better discrimination among isolates, being able to monitor the toxin genes possessed, and quantification, as well as species and strain attribution. Because molecular tools for identification and subtyping of *B. cereus s.l.* require the isolation of single strains (the pure culture of the Koch postulates), culture-based methods still remain an important step in food and clinical microbiology diagnostics. The development of new genomic typing tools, and moreover whole genome sequencing, opens future perspectives in diagnostics for food or clinical microbiology. These new tools open the way to identifying peculiar molecular characteristics of strains, both pathogenic and not; finding markers for new diagnostics methods; monitoring contamination routes; and tracing food-borne outbreaks by correlating patient isolates with food-derived isolates. However, this upgraded amount of information needs appropriate easy-to-use bioinformatic tools to render standard what is now of only academic use or interest.

## REFERENCES

1. Abshire TG, Brown JE, Ezzell JW. Production and validation of the use of gamma phage for identification of *Bacillus anthracis*. *J Clin Microbiol* 2005;**43**:4780–8.
2. Agaisse H, Gominet M, Okstad OA, Kolsto AB, Lereclus D. PlcR is a pleiotropic regulator of extracellular virulence factor gene expression in *Bacillus thuringiensis*. *J Clin Microbiol* 1999;**32**:1043–53.
3. Agren J, Sundström A, Håfström T, Segerman B. Gegenees: fragmented alignment of multiple genomes for determining phylogenomic distances and genetic signatures unique for specified target groups. *PLoS One* 2012;**7**:e39107. http://dx.doi.org/10.1371/journal.pone.0039107.

4. Anonymous. *Microbiology of food and animal feeding stuffs – horizontal method for the enumeration of presumptive Bacillus cereus – Colony-count technique at 30 degrees C.* ISO 7932:2004; 2004. TC 34/SC 9.
5. Bauer T, Stark T, Hofmann T, Ehling-Schulz M. Development of a stable isotope dilution analysis for the quantification of the *Bacillus cereus* toxin cereulide in foods. *J Agric Food Chem* 2010;**58**:1420–8.
6. Bottone EJ. *Bacillus cereus*, a volatile human pathogen. *Clin Microbiol Rev* 2010;**23**: 382–438.
7. Bravo A, Likitvivatanavong S, Gill SS, Soberón M. *Bacillus thuringiensis*: a story of a successful bioinsecticide. *Insect Biochem Mol Biol* 2011;**41**:423–31.
8. Cardazzo B, Negrisolo E, Carraro L, Alberghini L, Patarnello T, Giaccone V. Multiple-locus sequence typing and analysis of toxin genes in *Bacillus cereus* food-borne isolates. *Appl Environ Microbiol* 2008;**74**:850–60.
9. Carlson CR, Caugant DA, Kolstø AB. Genotypic diversity among *Bacillus cereus* and *Bacillus thuringiensis* strains. *Appl Environ Microbiol* 1994;**60**:1719–25.
10. Ceuppens S, Boon N, Rajkovic A, Heyndrickx M, VandeWiele T, Uyttendaele M. Quantification methods for *Bacillus cereus* vegetative cells and spores in the gastrointestinal environment. *J Microbiol Methods* 2010;**83**:202–10.
11. Chaves JQ, Pires ES, Vivoni AM. Genetic diversity, antimicrobial resistance and toxigenic profiles of *Bacillus cereus* isolated from food in Brazil over three decades. *Int J Food Microbiol* 2011;**147**:12–6.
12. Chen ML, Tsen HY. Discrimination of *Bacillus cereus* and *Bacillus thuringiensis* with 16 S rRNA and *gyrB* gene based PCR primers and sequencing of their annealing sites. *J Appl Microbiol* 2002;**92**:912–9.
13. Chon JW, Kim JH, Lee SJ, Hyeon JY, Seo KH. Toxin profile, antibiotic resistance and phenotypic and molecular characterization of *Bacillus cereus* in Sunsik. *Food Microbiol* 2012;**32**:217–22.
14. Daffonchio D, Borin S, Frova G, Gallo R, Mori E, Fani R, et al. A randomly amplified poly morphic DNA marker specific for the *Bacillus cereus* group is diagnostic for *Bacillus anthracis*. *Appl Environ Microbiol* 1999;**65**:1298–303.
15. Daffonchio D, Cherif A, Borin S. Homoduplex and heteroduplex polymorphisms of the amplified ribosomal 16S-23S internal transcribed spacers describe genetic relationships in the "*Bacillus cereus* group". *Appl Environ Microbiol* 2000;**66**:5460–8.
16. De Jonghe V, Coorevits A, De Block J, Van Coillie E, Grijspeerdt K, Herman L, et al. Toxinogenic and spoilage potential of aerobic spore-formers isolated from raw milk. *Int J Food Microbiol* 2010;**136**:318–25.
17. Demirev PA, Fenselau C. Mass spectrometry in biodefense. *J Mass Spectrom* 2008;**43**(11): 1441–57.
18. Dzieciol M, Fricker M, Wagner M, Hein I, Ehling-Schulz M. A diagnostic real-time PCR assay for quantification and differentiation of emetic and non-emetic *Bacillus cereus* in milk. *Food Control* 2013;**32**:176–85.
19. Ehling-Schulz M, Fricker M, Scherer S. Identification of emetic toxin producing *Bacillus cereus* strains by a novel molecular assay. *FEMS Microbiol Lett* 2004;**232**:189–95.
20. Ehling-Schulz M, Vukov N, Schulz A, Shaheen R, Andersson M, Märtlbauer E, et al. Identification and partial characterization of the nonribosomal peptide synthetase gene responsible for cereulide production in emetic *Bacillus cereus*. *Appl Environ Microbiol* 2005;**71**:105–13.
21. Ehling-Schulz M, Svensson B, Guinebretiere MH, Lindbäck T, Andersson M, Schulz A, et al. Emetic toxin formation of *Bacillus cereus* is restricted to a single evolutionary lineage of closely related strains. *Microbiology* 2005;**151**:183–97.
22. Ehling-Schulz M, Guinebretiere MH, Monthan A, Berge O, Fricker M, Svensson B. Toxin gene profiling of enterotoxic and emetic *Bacillus cereus*. *FEMS Microbiol Lett* 2006;**260**:232–40.

23. Ehling-Schulz M, Knutsson R, Scherer S. *Bacillus cereus*. In: Kathariou S, Fratamico P, Liu Y, editors. *Genomes of food- and water-borne pathogens.* Washington (DC): ASM Press; 2011. p. 147–64.

24. Ehling-Schulz M, Messelhausser U. Bacillus "next generation" diagnostics: moving from detection toward subtyping and risk-related strain profiling. *Front Microbiol* February 2013;**22**:4–32. http://dx.doi.org/10.3389/fmicb.2013.00032. eCollection 2013.

25. Elhanany E, Barak R, Fisher M, Kobiler D, Altboum Z. Detection of specific *Bacillus anthracis* spore biomarkers by matrix-assisted laser desorption/ionization time-of-flight mass spectrometry. *Rapid Commun Mass Spectrom* 2001;**15**(22):2110–6.

26. Enright MC, Spratt BG. Multilocus sequence typing. *Trends Microbiol* 1999;**7**:482–7.

27. Felder KM, Hoelzle K, Wittenbrink MM, Zeder M, Ehricht R, Hoelzle LE. A DNA microarray facilitates the diagnosis of *Bacillus anthracis* in environmental samples. *Lett Appl Microbiol* September 2009;**49**(3):324–31.

28. Fricker M, Messelhäußer U, Busch U, Scherer S, Ehling-Schulz M. Diagnostic real-time PCR assays for the detection of emetic *Bacillus cereus* strains in foods and recent food borne outbreaks. *Appl Environ Microbiol* 2007;**73**:1892–8.

29. Fricker M, Reissbrodt R, Ehling-Schulz M. Evaluation of standard and new chromogenic selective plating media for isolation and identification of *Bacillus cereus*. *Int J Food Microbiol* 2008;**121**:27–34.

30. Fricker M, Ågren J, Segerman B, Knutsson R, Ehling-Schulz M. Evaluation of *Bacillus* strains as model systems for the work on *Bacillus anthracis* spores. *Int J Food Microbiol* 2011;**145**:S129–36.

31. *Foodborne pathogenic microorganisms and natural toxins handbook Bacillus cereus and other Bacillus spp.* U.S. Food & Drug Administration (C.F.S.A.N); 2003.

32. Ghelardi E, Celandroni F, Salvetti S, Fiscarelli E, Senesi S. *Bacillus thuringiensis* pulmonary infection: critical role for bacterial membrane-demaging toxins and host neutrophils. *Microbes Infect* 2007;**9**:591–8.

33. Gordon RE, Haynes WC, Hor-Nay Pang C. *The genus Bacillus.* Agriculture Handbook No. 427. Washington (DC): USDA; 1973.

34. Guinebretière MH, Broussolle V, Nguyen-The C. Enterotoxigenic profiles of food-poisoning and food-borne *Bacillus cereus* strains. *J Clin Microbiol* 2002;**40**:3053–6.

35. Guinebretière MH, Thompson FL, Sorokin A, Normand P, Dawyndt P, Ehling-Schulz M, et al. Ecological diversification in the *Bacillus cereus* group. *Environ Microbiol* 2008;**10**:851–65.

36. Guinebretière MH, Velge P, Couvert O, Carlin F, Debuyser ML, Nguyen-The C. Ability of *Bacillus cereus* group strains to cause food poisoning varies according to phylogenetic affiliation (groups I to VII) rather than species affiliation. *J Clin Microbiol* 2010;**48**:3388–91.

37. Guinebretière MH, Auger S, Galleron N, Contzen M, De Sarrau B, De Buyser ML, et al. *Bacillus cytotoxicus* sp. nov. is a new thermotolerant species of the *Bacillus cereus* group occasionally associated with food poisoning. *Int J Syst Evol Microbiol* 2013;**6**:31–40.

38. Helgason E, Okstad OA, Caugant DA, Johansoen HA, Fouet A, et al. *Bacillus anthracis, Bacillus cereus* and *Bacillus thuringiensis* – one species on the basis of genetic evidence. *Appl Environ Microbiol* 2000;**66**:2627–30.

39. Hoffmaster AR, Novak RT, Marston CK, Gee JE, Helsel L, Pruckler JM, et al. Genetic diversity of clinical isolates of *Bacillus cereus* using multilocus sequence typing. *BMC Microbiol* 2008;**8**:191. http://dx.doi.org/10.1186/1471-2180-8-191.

40. Jackson SG, Goodbrand RB, Ahmed R, Kasatiya S. *Bacillus cereus* and *Bacillus thuringiensis* isolated in a gastroenteritis outbreak investigation. *Lett Appl Microbiol* 1995;**21**:103–5.

41. Jernigan JA, Stephens DS, Ashford DA, Omenaca C, Topiel MS, et al. Bioterrorism-related inhalational anthrax: the first 10 cases reported in the United States. *Emerg Infect Dis* 2001;**7**:933–44.

42. Jernigan JA, Stephens DS, Ashford DA, Omenaca C, Topiel MS, et al. Bioterrorism-related inhalational anthrax: investigation of bioterrorism-related anthrax, United States, 2001. Epidemiological findings. *Emerg Infect Dis* 2002;**8**:1019–28.

43. Keim P, Price LB, Klevytska AM, Smith KL, Schupp JM, Okinaka R, et al. Multiple-locus variable-number tandem repeat analysis reveals genetic relationships within *Bacillus anthracis*. *J Bacteriol* 2000;**182**:2928–36.

44. Kim W, Hong YP, Yoo JH, Lee WB, Choi CS, Chung SI. Genetic relationships of *Bacillus anthracis* and closely related species based on variable-number tandem repeat analysis and BOX-PCR genomic fingerprinting. *FEMS Microbiol Lett* 2002;**207**:21–7.

45. Klee SR, Ozel M, Appel B, Boesch C, Ellerbrok H, Jacob D, et al. Characterization of *Bacillus anthracis*-like bacteria isolated from wild great apes from Cote d'Ivoire and Cameroon. *J Bacteriol* 2006;**188**:5333–44.

46. Krishnamurthy T, Ross PL, Rajamani U. Detection of pathogenic and non-pathogenic bacteria by matrix-assisted laser desorption/ionization time-of-flight mass spectrometry. *Rapid Commun Mass Spectrom* 1996;**10**(8):883–8.

47. Koch R. Untersuchungen über Bakterien: V. Die Ätiologie der Milzbrand-Krakheit, begründet auf die Entwicklungsgeschichte des *Bacillus anthracis*. *Cohns Beitr Biol Pflanz* 1876;**2**:277–310.

48. Kolsto AB, Tourasse NJ, Okstad OA. What sets *Bacillus anthracis* apart from other *Bacillus* species? *Annu Rev Microbiol* 2009;**63**:451–76.

49. Kuroda M, Serizawa M, Okutani A, Sekizuka T, Banno S, Inoue S. Genome-wide single nucleotide polymorphism typing method for identification of *Bacillus anthracis* species and strains among *B. cereus* group species. *J Clin Microbiol* 2010;**48**:2821–9.

50. Lasch P, Beyer W, Nattermann H, et al. Identification of *Bacillus anthracis* by using matrix-assisted laser desorption ionization-time of flight mass spectrometry and artificial neural networks. *Appl Environ Microbiol* 2009;**75**(22):7229–42.

51. Leski TA, Caswell CC, Pawlowski M, Klinke DJ, Bujnicki JM, Hart SJ, et al. Identification and classification of *bcl* genes and proteins of *Bacillus cereus* group organisms and their application in *Bacillus anthracis* detection and fingerprinting. *Appl Environ Microbiol* 2009;**75**:7163–72.

52. Liu PY, Ke SC, Chen SL. Use of pulsed-field gel electrophoresis to investigate a pseudo-outbreak of *Bacillus cereus* in a pediatric unit. *J Clin Microbiol* 1997;**35**:1533–5.

53. Logan NA, Berkeley RCW. Identification of *Bacillus* strains using the API system. *J Gen Microbiol* 1984;**130**:1871–82.

54. Martínez-Blanch JF, Sánchez G, Garay E, Aznar R. Development of a real-time PCR assay for detection and quantification of enterotoxigenic members of *Bacillus cereus* group in food samples. *Int J Food Microbiol* 2009;**35**:15–21.

55. Mock M, Fouet A. Anthrax. *Annu Rev Microbiol* 2001;**55**:647–71.

56. Mullis KB, Faloona FA. Specific synthesis of DNA by polymerase-catalyzed chain reaction. *Methods Enzymol* 1987;**155**:335–50.

57. Orduz S, Restrepo W, Patino MM, Rojas W. Transfer of toxin genes to alternate bacterial host for mosquito control. *Mem Inst Oswaldo Cruz* 1995;**90**:97–107.

58. Ohsaki Y, Koyano S, Tachibana M, Shibukawa K, Kuroki M, Yoshida I, et al. Undetected *Bacillus* pseudo-outbreak after renovation work in a teaching hospital. *J Infect* 2007;**54**:617–22.

59. Peng H, Ford V, Frampton EW, Restaino L, Shelef LA, Spitz H. Isolation and enumeration of *Bacillus cereus* from foods on a novel chromogenic plating media. *Food Microbiol* 2001;**18**:231–8.

60. Phelps RJ, McKillip JL. Enterotoxin production in natural isolates of *Bacillaceae* outside the *Bacillus cereus* group. *Appl Environ Microbiol* 2002;**68**:3147–51.

61. Rasimus S, Mikkola R, Anderson MMA, Teplova VV, Venediktova N, Ek-Kommonen C, et al. Psychrotolerant *Paenibacillus tundrae* isolates from barley grains produce new cereulide-like depsipeptides (paenilide and homopaenilide) that are highly toxic to mammalian cells. *Appl Environ Microbiol* 2012;**78**:3732–43.

62. Raymond B, Wyres KL, Sheppard SK, Ellis RJ, Bonsall MB. Environmental factor determining the epidemiology and population genetic structure of the *Bacillus cereus* group in the field. *PLoS Pathog* 2010;**6**:e1000905. http://dx.doi.org/10.1371/journal.ppat.1000905.

63. Rasko DA, Altherr MR, Han CS, Ravel J. Genomics of the *Bacillus cereus* group of organisms. *FEMS Microbiol Lett* 2005;**29**:303–29.
64. Réjasse A, Gilois N, Barbosa I, Huillet E, Bevilacqua C, Tran S, et al. Temperature-dependent production of various PlcR-controlled virulence factors in *Bacillus weihenstephanensis* strain KBAB4. *Appl Environ Microbiol* 2012;**78**:2553–61.
65. Read TD, Turingan RS, Cook C, Giese H, Thomann UH, Hogan CC, et al. Rapid multi-locus sequence typing using microfluidic biochips. *PLoS One* 2010;**5**:e10595. http://dx.doi.org/10.1371/journal.pone.0010595.
66. Rowan NJ, Deans K, Anderson JG, Gemmell CG, Hunter IS, Chaitong T. Putative virulence factor expression by clinical and food isolates of *Bacillus* spp. after growth in reconstituted infant milk formulae. *Appl Environ Microbiol* 2001;**67**:3873–81.
67. Ryzhov V, Bundy JL, Fenselau C, Taranenko N, Doroshenko V, Prasad CR. Matrix-assisted laser desorption/ionization time-of-flight analysis of *Bacillus* spores using a 2.94 micron infrared laser. *Rapid Commun Mass Spectrom* 2000;**14**(18):1701–6.
68. Ryzhov V, Hathout Y, Fenselau C. Rapid characterization of spores of *Bacillus cereus* group bacteria by matrix-assisted laser desorption-ionization time-of-flight mass spectrometry. *Appl Environ Microbiol* 2000;**66**(9):3828–34.
69. Savini V, Favaro M, Fontana C, et al. *Bacillus cereus* heteroresistance to carbapenems in a cancer patient. *J Hosp Infect* 2009;**71**:288–90.
70. Segerman B, DeMedici D, Ehling-Schulz M, Fach P, Fenicia L, Fricker M, et al. Bioinformatic tools for using whole genome sequencing as a rapid high resolution diagnostic typing tool when tracing bioterror organisms in the food and feed chain. *Int J Food Microbiol* 2011;**145**:S167–76.
71. Senesi S, Celandroni F, Scher S, Wong ACL, Ghelardi E. Swarming motility in *Bacillus cereus* and characterization of a *fliY* mutant impaired in swarm cell differentiation. *Microbiology* 2002;**148**:1785–94.
72. Seng P, Drancourt M, Gouriet F, La Scola B, Fournier PE, Rolain JM, et al. Ongoing revolution in bacteriology: routine identification of bacteria by matrix-assisted laser desorption ionization time-of-flight mass spectrometry. *Clin Infect Dis* 2009;**49**:543–51.
73. Seng P, Rolain JJM, Fournier PE, La Scala B, Drancourt M, Raoult D. MALDI-TOF-mass spectrometry applications in clinical microbiology. *Future Microbiol* 2010;**5**(11):1733–54.
74. Seng P, Abat C, Rolain JM, Colson P, Lagier JC, Gouriet F, et al. Identification of rare pathogenic bacteria in a clinical microbiology laboratory: impact of Matrix-assisted laser desorption ionization-time of flight mass spectrometry. *J Clin Microbiol* 2013;**51**:2182–94.
75. Slamti L, Perchat S, Gominet M, Vilas-Boas G, Fouet A, et al. Distinct mutations in PlcR explain why some strains of the *Bacillus cereus* group are nonhemolytic. *J Bacteriol* 2004;**186**:3531–8.
76. Smith NR, Gordon RE, Clark FE. *Aerobic sporeforming bacteria*. Washington (DC): USDA; 1952. Monograph No. 16.
77. Stenfors Arnesen LP, Fagerlund A, Granum PE. From soil to gut: *Bacillus cereus* and its food poisoning toxins. *FEMS Microbiol Rev* 2008;**32**:579–606.
78. Stephan R. Randomly amplified polymorphic DNA (RAPD) assay for genomic fingerprinting of *Bacillus cereus* isolates. *Int J Food Microbiol* 1996;**31**:311–6.
79. Swaminathan B, Gerner-Smidt P, Ng LK, Lukinmaa S, Kam KM, Rolando S, et al. Building PulseNet International: an interconnected system of laboratory networks to facilitate timely public health recognition and response to foodborne disease outbreaks and emerging foodborne diseases. *Foodborne Pathog Dis* Spring 2006;**3**(1):36–50.
80. Tille MP, editor. *Bailey & Scott's diagnostic microbiology*. 13th ed. Mosby, Inc. Elsevier edition.
81. Thorsen L, Hansen BM, Nielsen KF, Hendriksen NB, Phipps RK, Budde BB. Characterization of emetic *Bacillus weihenstephanensis*, a new cereulide-producing bacterium. *Appl Environ Microbiol* 2006;**72**:5118–21.

82. Tourasse NJ, Helgason E, Økstad OA, Hegna IK, Kolstø AB. The *Bacillus cereus* group: novel aspects of population structure and genome dynamics. *J Appl Microbiol* 2006;**101**:579–93.

83. Tourasse NJ, Kolsto AB. SuperCAT: a super tree database for combined and integrative multilocus sequence typing analysis of the *Bacillus cereus* group of bacteria (including *B. cereus, B. anthracis* and *B. thuringiensis*). *Nucleic Acids Res* 2008;**36**:D461–8.

84. Tourasse NJ, Okstad OA, Kolstø AB. HyperCAT: an extension of the SuperCAT database for global multischeme and multidata type phylogenetic analysis of the *Bacillus cereus* group population. *Database (Oxford)* 2010;**2010**:baq017.

85. Tourasse NJ, Helgason E, Klevan A, Sylvestre P, Moya M, Haustant M, et al. Extended and global phylogenetic view of the *Bacillus cereus* group population by combination of MLST, AFLP, and MLEE genotyping data. *Food Microbiol* 2011;**28**:236–44.

86. Tsilia V, Devreese B, de Baenst I, Mesuere B, Rajkovic A, Uyttendaele M, et al. Application of MALDI-TOF mass spectrometry for the detection of enterotoxins produced by pathogenic strains of the *Bacillus cereus* group. *Anal Bioanal Chem* 2012;**404**:1691–702.

87. Turnbull PCB, Kramer J, Melling J. Bacillus. In: 8th ed. Topley WWC, Wilson GS, editors. *Topley and Wilson's principles of bacteriology, virology and immunity*, vol. 2. London (UK): Edward Arnold; 1990. p. 188–210.

88. Uchino Y, Iriyama N, Matsumoto K, et al. A case series of *Bacillus cereus* septicemia in patients with hematological disease. *Intern Med* 2012;**51**:2733–8.

89. Vassileva M, Torii K, Oshimoto M, Okamoto A, Agata N, Yamada K, et al. Phylogenetic analysis of *Bacillus cereus* isolates from severe systemic infections using multilocus sequence typing scheme. *Microbiol Immunol* 2006;**50**:743–9.

90. Wehrle E, Didier A, Moravek M, Dietrich R, Märtlbauer E. Detection of *Bacillus cereus* with enteropathogenic potential by multiplex real-time PCR based on SYBR Green I. *Mol Cell Probes* 2010;**24**:124–30.

91. Wielinga PR, Hamidjaja RA, Agren J, Knutsson R, Segermanm B, Fricker M, et al. A multiplex real-time PCR for identifying and differentiating *B. anthracis* virulent types. *Int J Food Microbiol* 2011;**145**:S137–44.

# CHAPTER 3

# *Bacillus cereus* Hemolysins and Other Virulence Factors

**Rosa Visiello[1], Stefano Colombo[2], Edoardo Carretto[1]**
[1]Clinical Microbiology Laboratory, IRCCS Arcispedale Santa Maria Nuova, Reggio Emilia, Italy;
[2]Mérieux NutriSciences, Lyon, France

## SUMMARY

*Bacillus cereus* is mainly known for causing food poisoning and severe non-gastrointestinal tract infections. The intestinal and nonintestinal pathogenicity of this microorganism is mainly due to the synergistic effects of a number of virulence products that promote intestinal cell destruction and/or resistance to the host immune system. The various substances produced by *B. cereus* are mainly enterotoxins, hemolysins, phospholipases and emetic toxin, whose activity may overlap in causing human disease. This review briefly describes the characteristics of the main virulence factors of *B. cereus*.

## *BACILLUS CEREUS* VIRULENCE FACTORS

*Bacillus cereus* is mainly known for causing food poisoning and severe eye infections, but it is also an opportunistic human pathogen, causing different local and systemic infections. In these cases, there is still poor recognition of the various mechanisms ruled by *B. cereus* in the pathogenesis of the human diseases. In early stationary phase, *B. cereus* produces several toxins, such as degradation enzymes, cytotoxic factors etc., which act in a synergistic way. These substances are produced mainly during bacterial growth and are summarized in Table 1.[1,2] Various molecules have activity against erythrocytes, either directly (hemolysins) or as a part of their activity. The emetic form of *B. cereus* food poisoning is attributed to one single toxin, cereulide. The diarrheal form is mainly due to five different enterotoxins and degenerative enzymes.

In this chapter, we briefly describe the activity of the enterotoxic complexes (hemolysin BL (Hbl) and nonhemolytic enterotoxin (Nhe)), the phospholipases-C/sphingomyelinase complex, the emetic toxin (cereulide), and the various hemolysins–cytolysins (types I–IV).

*The Diverse Faces of Bacillus cereus*
ISBN 978-0-12-801474-5
http://dx.doi.org/10.1016/B978-0-12-801474-5.00003-7

**Table 1** Toxins and other substances released by *Bacillus cereus*

| Toxin | Main characteristics |
|---|---|
| Hemolysin BL (Hbl) | Protein, three components: B, L1, and L2 |
| | Hemolysis, cytotoxicity, dermonecrosis, capillary permeability |
| Nonhemolytic enterotoxin (Nhe) | Protein, three components (39, 45, and 105 kDa) |
| Enterotoxin T (BceT) | Protein, one component 40/41 kDa |
| | Cytotoxicity, capillary permeability |
| Enterotoxin FM (EntFM) | Protein, one component, 45 (48) kDa |
| | Capillary permeability |
| Cytotoxin K (CytK) | Protein, one component, 34 kDa |
| | Hemolysis, cytotoxicity |
| Emetic toxin (cereulide) | Cyclopeptide, 1.2 kDa |
| Sphingomyelinase | 34 kDa, hemolytic, part of Nhe? |
| Phospholipase C | Degranulation of human neutrophils |
| Phosphatidylinositol hydrolase | Breakage of the protein anchorage on plasma membranes |
| Phosphatidylcholine hydrolase | General hydrolytic action |
| Hemolytic sphingomyelinase | Sphingomyelin harm |
| Hemolysin I (cereolysin O) | Thiol-dependent, thermolabile, inhibited by cholesterol |
| Hemolysin II | Thermolabile, not inhibited by cholesterol |
| Hemolysin III | Formation of transmembrane pores |

Modified from Ref. 1.

The **hemolysin BL (Hbl)** is the enterotoxin that can causes diarrhea; it is produced by 60% of *B. cereus* strains and is also produced by *Bacillus thuringiensis* and *Bacillus mycoides*. It consists of three fractions transcribed from the same operon, *hbl*, located on the stretch of unstable chromosome in *B. cereus*. Hbl expresses dermonecrotic activity and also has the ability to alter the permeability of blood vessels, causing an accumulation of fluid in the intraluminal ileal bend. The three components of Hbl, fraction B (binding component) and lytic components L1 and L2, bind to the target cell independently and reversibly, forming a complex causing a lytic action toward the cellular membrane by forming pores in it. All three Hbl components are required for biological activity. One of the hypotheses is that once bound, the various components join to form a single complex, forming transmembrane pores consisting of at least one of each component, but specific membrane receptors have not been identified. A high degree of molecular heterogeneity exists in Hbl toxin between strains. According to a study on 127 isolates of *B. cereus*, four different sizes of fraction B (38, 42, 44, and 46 kDa), three of L2 (43, 45, and 49 kDa), and two of L1 (38 and 41 kDa), were identified using western blot analysis, with individual strains producing varying combinations of single and multiple bands of each component and a total of 13 different band patterns being observed.[3] The crystal structure of Hbl has some homologies with the pore-forming hemolysin called cytolysin A from *Escherichia coli*.[4]

**Nonhemolytic enterotoxin (Nhe)** is almost always expressed by *B. cereus* strains and thus is considered one of its main virulence factors. Nhe is a three-component toxin, which in binary combination exhibits lower pathogenic potential compared to the complete toxin. It is transcribed by the operon *nhe*, consisting of three open reading frames (*nheA*, *nheB*, and *nheC*). The first two transcribe two components of 45 and 39 kDa, respectively, while the third (105 kDa) has a function not fully understood, but considered necessary for the final synthesis of the toxin.[1] Nhe exposure leads to a rapid membrane lysis by forming pores in lipid bilayers, resulting in a colloido-osmotic lysis. The cytotoxic activity of Nhe toward epithelial cells is that of rapid disruption of the plasma membrane following exposure and formation of pores in planar lipid bilayers.[4] As indicated in its name, in the past it was supposed that this enterotoxin had no hemolytic activity. However, a 2008 study pointed out that Nhe is somewhat able to induce hemolysis in erythrocytes from some mammalian species.[4]

The **enterotoxin T (BceT)** is a 40/41-kDa enterotoxin that is encoded by the *bceT* gene. It is released during cellular lysis. Even when *bceT* has been cloned into *E. coli*, no biological activity is detected, either in the supernatant or in the cell extract, thus suggesting that its role in effecting the symptoms of food poisoning is still a matter of debate.[5]

Little or nothing is known about the characteristics of **enterotoxin FM (EntFM)**. It is a cell wall peptidase (also known as CwpFM) that seems to play a role in bacterial shape, motility, adhesion potential to epithelial cells, biofilm formation, and vacuolization of macrophages.[6]

The **cytotoxin K (CytK)**, also called hemolysin IV, is a 34-kDa protein that expresses both cytotoxic and hemolytic properties. CytK is quite different from Hbl and Nhe, being characterized by one component. First described by Beecher et al.,[7] it has been subsequently characterized by Lund et al.[8] from *B. cereus* strains that caused a severe outbreak; these isolates seemed to express only this kind of virulence factor. The deduced amino acid sequence of the toxin showed similarity to *Staphylococcus aureus* leukocidins (γ-hemolysin and α-hemolysin, sharing 30% homology with the latter), to *Clostridium perfringens* β-toxin, and to another *B. cereus* hemolysin (type II, 37% homologies). All these proteins belong to the family of β-barrel channel-forming toxins.[2,8] Subsequently, a variant of this toxin, sharing 89% similarity to the amino acid sequence of the first CytK isolated by Lund et al., has been isolated and named as CytK-2.[9] Further studies performed on these cytotoxins allowed for the taxonomical discrimination of a new *Bacillus* species: the strains harboring CytK-1 have been currently defined as *Bacillus cytotoxicus*, which is de facto considered different from the *B. cereus* strains harboring the CytK-2 variant.[10]

The mechanism of action for CytK is not completely understood. It is known that this toxin is able to produce pores in the target cells, which are mainly epithelial cells. These proteins are secreted in a soluble form, being then converted into a transmembrane pore by the assembly of an oligomeric β-barrel, with the hydrophobic residues facing the lipids and the hydrophilic residues facing the lumen of the channel.[2]

Hbl, Nhe, and CytK express maximum activity during the end of the exponential phase or the beginning of the stationary phase, with operons being regulated by the gene *plcR* (phospholipase C regulator), which also regulates the expression of *plcA* for phospholipase C activity and belongs to the PlcR/PapR quorum-sensing system. The promoter region of PlcR-regulated genes has a highly conserved palindromic region that acts as a specific target for PlcR activation.[2,11] The gene transcription starts usually at the onset of the stationary phase and reaches maximum expression in a few hours. A 2013 study performed on the *B. cereus* ATCC-14579 strain pointed out that only a small subpopulation is responsible for CytK production in a homogeneous monoculture.[12] However, environmental factors such as pH, growth rate, and the amount of oxygen and carbohydrates all further regulate and have an impact on toxin production.[13]

For its activity, PlcR requires a peptide expressed as a propeptide (PapR) under the control of PlcR, which is exported out of the cell, processed to generate an active heptapeptide, and then internalized into the bacterial cell.[2,14,15] The PlcR/PapR system is now known to be the central transcriptional regulator for virulence genes in *B. cereus* at the onset of the stationary phase, controlling the expression of at least 45 genes, mainly virulence factors.[2,11]

Some papers seem to indicate that CytK production is characteristic of strains causing diarrheal illness rather than systemic diseases.[16] However, in another study, which was primarily focused on the performance evaluation of a multiplex polymerase chain reaction to rapidly detect various *B. cereus* genes, the *cytK* gene was detected in 88.8% of *B. cereus* and in 83.9% of

*B. thuringiensis* strains isolated from different sources, irrespective of their causative role in human diseases.[17]

The **emetic toxin (or factor or emetic cereulide)**, is a small protein (1.2 kDa), which is nonimmunogenic although it possesses immunomodulating properties. It acts on human natural killer (NK) cells by inhibiting cytotoxicity and cytokine production, as well as causing mitochondrial swelling, resulting in possible apoptosis of NK cells.[18,19] This protein is highly heat stable (126 °C for 90 min), also being stable under extreme pH conditions (range: 2–11). It appears to be insensitive to enzymatic digestion, with no human enzymes known to be able to inactivate it. It is produced during the bacterial stationary phase in contaminated food, for example, in a dish made of rice maintained at 25–30 °C, with the relationship to the sporulation of the bacterial cells remaining unclear.[1] Its structure was unknown for a long time, until the discovery of its vacuolating action on Hep-2 cells, which allowed the protein purification and the evaluation of the structural details. In particular, it is organized in a three-ring configuration, each ring of which contains four amino acids and/or oxyacids (dodecadepsipeptide),[20,21] similar to valinomycin, a potassium ionophore.[22] The toxin is probably a peptide-synthetic enzyme, produced by a nonribosomal peptide synthase, encoded by the 24-kb cereulide synthase (*ces*) gene cluster. Whereas all the enterotoxins are located chromosomally, the cereulide synthase genes are located on megaplasmids with a pXO1-like backbone.[17,23] Also, because it acts as a cation ionophore, it has the ability to inhibit mitochondrial activity, which can lead to serious degeneration of target cells. Although the exact mechanisms regulating the cereulide production are not fully characterized, factors such as oxygen, temperature, and pH affect its expression, which is regulated by the transitional state regulator ArbB and not the PlcR.

In vivo, the mechanism of action is due to the stimulation of vagal afferents through specific binding of the $5\text{-HT}_3$ receptors. Emesis induced by cereulide occurs in minutes following oral administration, thus indicating rapid absorption of the substance from the esophagus into the neuronal serotonin receptors. This is possibly attributable to its lipophilic properties, which are similar to dietary lipids, leading to good penetration into cells, and thus high concentrations may be found even in blood following ingestion.[19]

The mitochondrial swelling in target cells may be a direct consequence of the ionophoric action of cereulide, as mitochondrial $K^+$ channels are known to play an influential role in mitochondrial volume control. These peptides create $K^+$ channels, which result in an efflux of $K^+$ and rapid but transient hyperpolarization of the membrane, resulting in cell damage and subsequent apoptosis with the activation of specific caspases.[19]

Lesions due to the actions of toxins are nonspecific and usually result in mucus-catarrhal forms, sometimes fibrin-catarrhal. Hepatic degeneration and necrotic–hemorrhagic enteritis are the most serious of injuries that follow the action of cereulide and CytK, respectively.

*Bacillus cereus* also secretes membrane–damaging toxins, **phospholipases**, such as *Bc*-SMase (*B. cereus* sphingomyelinase), phosphatidylinositol-specific phospholipase C, and phosphatidylcholine-specific phospholipase C (PC-PLC). In a study using mouse macrophages, treatment with *Bc*-SMase resulted in a reduction in the generation of $H_2O_2$ and phagocytosis of macrophages induced by peptidoglycan, suggesting that *Bc*-SMase is essential for the hydrolysis of sphingomyelinase in macrophage membranes, which leads to a reduction in phagocytosis, therefore evading the immune response. Moreover, *Bc*-SMase appeared to cause hemolysis of sheep erythrocytes.[24]

*Bacillus cereus* may also cause hemolysis through the synergistic action of PC-PLC and *Bc*-SMase, which form a biological complex known as cereolysin AB.[30] This complex specifically hydrolyzes sphingomyelin in the intact erythrocyte membranes, leading to their disruption and subsequent hemolysis.[25]

**Hemolysin I (HlyI)** is also known as cereolysin O; very similar proteins are produced by *B. thuringiensis* and *Bacillus anthracis* (>98% similarity).[2] HlyI is a heat-labile protein that is inhibited by cholesterol, belonging to the cholesterol-dependent cytolysin (CDC) family (formerly known as thiol-activated cytolysins), which also comprises the pneumolysin produced by *Streptococcus pneumoniae*, streptolysin O produced by *Streptococcus pyogenes*, and listeriolysin produced by *Listeria monocytogenes*, which explains why there is a cross-reaction in immunodiffusion tests.[26]

The CDC family members share a common mechanism of action: these proteins disrupt the outer cell membrane through the formation of pores, leading to the destruction of the cytoplasmic membrane of erythrocytes of various hosts. For its action, HlyI requires only the presence of cholesterol in the target membrane, without specific receptors. Pore forming by CDCs follows the oligomerization and assembly of soluble monomers into a ring-shaped prepore, which undergoes a conformational change for insertion into the membrane, to form a large amphipathic transmembrane β-barrel structure.[2] The structure and molecular mechanism of several other CDCs are now also relatively well characterized and well described by Ramarao and Sanchis.[2]

Since CDCs are cytolytic proteins, they can lyse or permeabilize various types of host cell or intracellular organelles during infection. Therefore, *B. cereus* HlyI can play multiple roles during bacterial infections.

As for Hbl, Nhe, and CytK, the expression of HlyI is under the control of the transcriptional activator PlcR, which was first linked to *plcA*, the gene encoding phosphatidylinositol-specific phospholipase C.[2]

**Hemolysin II** (HlyII) is a thermolabile protein that is not inhibited by cholesterol and is degraded by proteolytic enzymes. It is produced as a larger preprotein, which is then reduced to its final form of 42.6 kDa.[27] HlyII is an oligomeric β-barrel pore-forming toxin with some kind of similarity in its amino acid sequence to the *S. aureus* α-toxin (being, however, 15 times more potent than the latter on rabbit blood cells),[28] to the β-toxin of *C. perfringens*, and to CytK.[29] HlyII binds to membranes through a 94-amino-acid C-terminal extension and finally inserts a glycine-rich segment into the membrane to form the walls of a transmembrane pore. Contemporarily, HlyII has been shown to produce anion-selective channels in planar lipid bilayers.[28–30] For its action, HlyII does not require neither the presence of cholesterol in the target membrane nor specific receptors; it can induce lysis in the target cells, which are a broad range of erythrocytes and phagocytic cells, but not in epithelial cells.[31]

As for other virulence factors herein described, HlyII plays different roles in *B. cereus* virulence. However, it is not directly involved in the diarrheal phenomenon, maybe because it is susceptible to trypsin digestion (which can happen in the small intestine), being in this sense similar to the β-toxin of *C. perfringens*, even if this is still hypothetical.[2] On the other hand, it is clearly established that HlyII causes cell death by apoptosis,[31] and in a paper by Cadot et al. HlyII was demonstrated, using quantitative molecular techniques, to be largely present in clinical strains isolated during human infections, allowing them to postulate that the *hlyII* gene could be carried only by strains with a pathogenic potential.[32] Moreover, it acts also in allowing the persistence and dissemination of *B. cereus* in the host through its ability to induce apoptosis of host monocytes and macrophages in vivo.[31]

HlyII is one of the few substances of *B. cereus* that is not regulated by PlcR. Instead, it is downregulated by the specific transcriptional regulator HlyIIR[33,34] and by the ferric uptake regulator (Fur). In particular, glucose 6P seems to directly bind to HlyIIR, enhancing in this way its activity at a posttranscriptional level, acting then on the *hlyII* gene, inhibiting its expression, which is finally modulated by the availability of glucose. On the other hand, *hlyII* is downregulated by iron during bacterial infections. HlyII expression is negatively regulated by iron via Fur by direct interaction with the *hlyII* promoter. DNase I footprinting and in vitro transcription studies indicate that Fur prevents RNA polymerase binding to the *hlyII* promoter. Both HlyIIR and Fur regulate *hlyII* expression in a concerted fashion, with the effect of Fur being maximal in the early stages of bacterial growth.[35] A very intriguing model of pathogenetic moment is suggested by Ramarao and Sanchis, who postulate that, when glucose is consumed by the bacteria and iron is sequestered by phagocytic cells as a natural host defense, the HlyIIR and Fur repressors become inactivated and *hlyII* expression is triggered. HlyII is then produced by the bacteria and secreted, triggering the death of hemocytes and macrophages.[2]

**Hemolysin III (HlyIII)** is the least characterized hemolytic toxin from the *B. cereus* group. It is a heat-labile protein that has never been purified but is encoded by a 657-nucleotide gene characterized in *E. coli*. Its hemolytic activity is not inhibited by cholesterol.[2]

The role of this hemolysin has not been investigated in vivo and can only be postulated. It is known that its mechanism of action is based on the formation of pores on the erythrocyte surface. It is likely that this toxin forms transmembrane pores in three steps: protein binding to the erythrocyte surface, monomer assembly to form the pores, and subsequent erythrocyte lysis.[2,36]

## REFERENCES

1. Colombo S, Carretto E. *Bacillus cereus* e specie correlate. In: Rondanelli EG, Fabbi M, Marone P, editors. *Trattato sulle infezioni e tossinfezioni alimentari*. Pavia (Italy): Selecta Medica; 2005. p. 609–39.
2. Ramarao N, Sanchis V. The pore-forming haemolysins of *Bacillus cereus*: a review. *Toxins* 2013;**5**(6):1119–39.
3. Kumar TD, Murali HS, Batra HV. Construction of a non toxic chimeric protein (L1-L2-B) of Haemolysin BL from *Bacillus cereus* and its application in HBL toxin detection. *J Microbiol Methods* 2008;**75**(3):472–7.
4. Fagerlund A, Lindback T, Storset AK, Granum PE, Hardy SP. *Bacillus cereus* Nhe is a pore-forming toxin with structural and functional properties similar to the ClyA (HlyE, SheA) family of haemolysins, able to induce osmotic lysis in epithelia. *Microbiology* 2008;**154**(Pt 3):693–704.
5. Choma C, Granum PE. The enterotoxin T (BcET) from *Bacillus cereus* can probably not contribute to food poisoning. *FEMS Microbiol Lett* 2002;**217**(1):115–9.
6. Tran SL, Guillemet E, Gohar M, Lereclus D, Ramarao N. CwpFM (EntFM) is a *Bacillus cereus* potential cell wall peptidase implicated in adhesion, biofilm formation, and virulence. *J Bacteriol* 2010;**192**(10):2638–42.

7. Beecher DJ, Wong AC. Cooperative, synergistic and antagonistic haemolytic interactions between haemolysin BL, phosphatidylcholine phospholipase C and sphingomyelinase from *Bacillus cereus*. *Microbiology* 2000;**146**(Pt 12):3033–9.

8. Lund T, De Buyser ML, Granum PE. A new cytotoxin from *Bacillus cereus* that may cause necrotic enteritis. *Mol Microbiol* 2000;**38**(2):254–61.

9. Fagerlund A, Ween O, Lund T, Hardy SP, Granum PE. Genetic and functional analysis of the cytK family of genes in *Bacillus cereus*. *Microbiology* 2004;**150**(Pt 8):2689–97.

10. Guinebretiere MH, Auger S, Galleron N, Contzen M, De Sarrau B, De Buyser ML, et al. *Bacillus cytotoxicus* sp. nov. is a novel thermotolerant species of the *Bacillus cereus* group occasionally associated with food poisoning. *Int J Syst Evol Microbiol* 2013;**63**(Pt 1):31–40.

11. Gohar M, Faegri K, Perchat S, Ravnum S, Okstad OA, Gominet M, et al. The PlcR virulence regulon of *Bacillus cereus*. *PLoS One* 2008;**3**(7):e2793.

12. Ceuppens S, Timmery S, Mahillon J, Uyttendaele M, Boon N. Small *Bacillus cereus* ATCC 14579 subpopulations are responsible for cytotoxin K production. *J Appl Microbiol* 2013;**114**(3):899–906.

13. Slamti L, Perchat S, Huillet E, Lereclus D. Quorum sensing in *Bacillus thuringiensis* is required for completion of a full infectious cycle in the insect. *Toxins* 2014;**6**(8):2239–55.

14. Gominet M, Slamti L, Gilois N, Rose M, Lereclus D. Oligopeptide permease is required for expression of the *Bacillus thuringiensis* plcR regulon and for virulence. *Mol Microbiol* 2001;**40**(4):963–75.

15. Bouillaut L, Perchat S, Arold S, Zorrilla S, Slamti L, Henry C, et al. Molecular basis for group-specific activation of the virulence regulator PlcR by PapR heptapeptides. *Nucleic Acids Res* 2008;**36**(11):3791–801.

16. Guinebretiere MH, Broussolle V, Nguyen-The C. Enterotoxigenic profiles of food-poisoning and food-borne *Bacillus cereus* strains. *J Clin Microbiol* 2002;**40**(8):3053–6.

17. Ngamwongsatit P, Buasri W, Pianariyanon P, Pulsrikarn C, Ohba M, Assavanig A, et al. Broad distribution of enterotoxin genes (hblCDA, nheABC, cytK, and entFM) among *Bacillus thuringiensis* and *Bacillus cereus* as shown by novel primers. *Int J Food Microbiol* 2008;**121**(3):352–6.

18. Agata N, Ohta M, Yokoyama K. Production of *Bacillus cereus* emetic toxin (cereulide) in various foods. *Int J Food Microbiol* 2002;**73**(1):23–7.

19. Paananen A, Mikkola R, Sareneva T, Matikainen S, Hess M, Andersson M, et al. Inhibition of human natural killer cell activity by cereulide, an emetic toxin from *Bacillus cereus*. *Clin Exp Immunol* 2002;**129**(3):420–8.

20. Agata N, Mori M, Ohta M, Suwan S, Ohtani I, Isobe M. A novel dodecadepsipeptide, cereulide, isolated from *Bacillus cereus* causes vacuole formation in HEp-2 cells. *FEMS Microbiol Lett* 1994;**121**(1):31–4.

21. Ehling-Schulz M, Svensson B, Guinebretiere MH, Lindback T, Andersson M, Schulz A, et al. Emetic toxin formation of *Bacillus cereus* is restricted to a single evolutionary lineage of closely related strains. *Microbiology* 2005;**151**(Pt 1):183–97.

22. Mikkola R, Saris NE, Grigoriev PA, Andersson MA, Salkinoja-Salonen MS. Ionophoretic properties and mitochondrial effects of cereulide: the emetic toxin of *B. cereus*. *Eur J Biochem/FEBS* 1999;**263**(1):112–7.

23. Ehling-Schulz M, Guinebretiere MH, Monthan A, Berge O, Fricker M, Svensson B. Toxin gene profiling of enterotoxic and emetic *Bacillus cereus*. *FEMS Microbiol Lett* 2006;**260**(2):232–40.

24. Oda M, Takahashi M, Matsuno T, Uoo K, Nagahama M, Sakurai J. Hemolysis induced by *Bacillus cereus* sphingomyelinase. *Biochim Biophys Acta* 2010;**1798**(6):1073–80.

25. Gilmore MS, Cruz-Rodz AL, Leimeister-Wachter M, Kreft J, Goebel W. A *Bacillus cereus* cytolytic determinant, cereolysin AB, which comprises the phospholipase C and sphingomyelinase genes: nucleotide sequence and genetic linkage. *J Bacteriol* 1989;**171**(2):744–53.

26. Kreft J, Berger H, Hartlein M, Muller B, Weidinger G, Goebel W. Cloning and expression in *Escherichia coli* and *Bacillus subtilis* of the hemolysin (cereolysin) determinant from *Bacillus cereus. J Bacteriol* 1983;**155**(2):681–9.

27. Baida G, Budarina ZI, Kuzmin NP, Solonin AS. Complete nucleotide sequence and molecular characterization of hemolysin II gene from *Bacillus cereus. FEMS Microbiol Lett* 1999;**180**(1):7–14.

28. Miles G, Bayley H, Cheley S. Properties of *Bacillus cereus* hemolysin II: a heptameric transmembrane pore. *Protein Sci* 2002;**11**(7):1813–24.

29. Andreeva ZI, Nesterenko VF, Yurkov IS, Budarina ZI, Sineva EV, Solonin AS. Purification and cytotoxic properties of *Bacillus cereus* hemolysin II. *Protein Expr Purif* 2006;**47**(1):186–93.

30. Andreeva ZI, Nesterenko VF, Fomkina MG, Ternovsky VI, Suzina NE, Bakulina AY, et al. The properties of *Bacillus cereus* hemolysin II pores depend on environmental conditions. *Biochim Biophys Acta* 2007;**1768**(2):253–63.

31. Tran SL, Guillemet E, Ngo-Camus M, Clybouw C, Puhar A, Moris A, et al. Haemolysin II is a *Bacillus cereus* virulence factor that induces apoptosis of macrophages. *Cell Microbiol* 2011;**13**(1):92–108.

32. Aboul-Nasr MB, Obied-Allah MR. Biological and chemical detection of fumonisins produced on agar medium by *Fusarium verticillioides* isolates collected from corn in Sohag, Egypt. *Microbiology* 2013;**159**(Pt 8):1720–4.

33. Budarina ZI, Nikitin DV, Zenkin N, Zakharova M, Semenova E, Shlyapnikov MG, et al. A new *Bacillus cereus* DNA-binding protein, HlyIIR, negatively regulates expression of *B. cereus* haemolysin II. *Microbiology* 2004;**150**(Pt 11):3691–701.

34. Guillemet E, Tran SL, Cadot C, Rognan D, Lereclus D, Ramarao N. Glucose 6P binds and activates HlyIIR to repress *Bacillus cereus* haemolysin *hlyII* gene expression. *PLoS One* 2013;**8**(2):e55085.

35. Sineva E, Shadrin A, Rodikova EA, Andreeva-Kovalevskaya ZI, Protsenko AS, Mayorov SG, et al. Iron regulates expression of *Bacillus cereus* hemolysin II via global regulator Fur. *J Bacteriol* 2012;**194**(13):3327–35.

36. Baida GE, Kuzmin NP. Mechanism of action of hemolysin III from *Bacillus cereus. Biochim Biophys Acta* 1996;**1284**(2):122–4.

# CHAPTER 4

# *Bacillus cereus* Mechanisms of Resistance to Food Processing

**Sarah M. Markland, Dallas G. Hoover**
Department of Animal and Food Sciences, University of Delaware, Newark, DE, USA

## SUMMARY

It is estimated that there are an average of 63,623 illnesses and 20 hospitalizations caused by *Bacillus cereus* every year in the United States. Spores of *Bacillus* species are of concern in the food industry, owing to their common occurrence and resistance to food processing methods. Common foods associated with contamination by *B. cereus* include rice, milk, grains, cereals, potatoes, vegetables, and low-nutrient foods. Evolving species, such as psychrotolerant strains of *B. cereus*, are of even greater concern because they have adapted to growth at reduced temperatures.

## INTRODUCTION

Since 1995 there has been a significant increase in demand by consumers for convenient food products of high quality, which are commonly referred to as ready-to-eat, cooked or chilled, or refrigerated processed foods of extended durability.[1] Mild processing techniques that are thought to help preserve the organoleptic properties of foods can subsequently allow spore-forming bacteria such as *Bacillus* and *Clostridium* species to survive in the food product.[2] Pathogenic *Bacillus* species may have the ability not only to germinate and grow at refrigeration temperatures[2,3] but also to produce toxins in foods. Evolving species, such as psychrotolerant strains of *B. cereus*, are of even greater concern because they have adapted to growth at reduced temperatures.

Spore formers can serve as spoilage agents in foods, and pathogenic types exist. Spore-forming bacteria can be problematic in food products for which mild processing is used,[4] owing to their common occurrence and extreme resistance to food processing methods, including heat, pressure, acidification, desiccation, and chemical disinfectants.[5–7] Spore formation is

*The Diverse Faces of Bacillus cereus*
ISBN 978-0-12-801474-5
http://dx.doi.org/10.1016/B978-0-12-801474-5.00004-9

triggered by nutrient depletion whereby a vegetative cell enters dormancy, but the spore is still able to respond to various agents in its environment, including temperature, pH, and the presence of nutrients, and germinate to resume metabolic activity.[5,8]

## BACILLUS CEREUS AND CHALLENGES FOR THE FOOD INDUSTRY

### The Emergence of Psychrotolerant *Bacillus cereus* and *Bacillus weihenstephanensis*

There is a clear correlation between the phylogenic groups of *B. cereus* and their adaptation to temperature, pH, and water activity. These properties can be used to predict the risk of a particular species to cause food-borne illness.[9] Several epidemiological studies have been performed comparing the genetic sequences of a variety environmental and food isolates of *B. cereus*. Psychrotolerant *B. cereus* isolates are more genetically similar to other psychrotolerant species, including *Bacillus mycoides*, than to mesophilic *B. cereus* strains, which are more genetically similar to *Bacillus thuringiensis*.[10–13] Therefore, it has been suggested that the taxonomy of the *B. cereus* group be revised. In 1998, a new species named *B. weihenstephanensis* was proposed to accommodate the psychrotolerant strains of *B. cereus*.[14]

Because of their potential to grow in refrigerated food products, their ability to produce toxins, and their implications in food-borne outbreaks, psychrotolerant species of *B. cereus*, including *B. weihenstephanensis*, have been of concern in the food industry.[15] *Bacillus weihenstephanensis* is a known causative agent of spoilage in white liquid egg products but can also cause spoilage of pasteurized milk. The *B. weihenstephanensis* strain isolated from a spoiled whole liquid egg product also demonstrated the ability to adhere to surfaces and form biofilms. These films can form on processing equipment commonly used in egg-breaking facilities, including stainless steel, model hydrophilic materials (glass), and model hydrophobic materials (polytetrafluoroethylene).[15]

### Minimally Processed Foods

The increase in consumer demand for convenient food products of premium sensory quality, including ready-to-eat, cooked, or chilled foods, and minimally processed foods, has led to the development of food products known as refrigerated processed foods of extended durability or RPFEDs.[1] These products are normally low-acid foods refrigerated at temperatures close to freezing of the food to maintain freshness and safety (i.e., a national cold chain); these foods are often globally sourced. Fluid products usually are pasteurized. Produce and other raw foods commonly rely on surface cleaning/washing, modified or controlled atmosphere packaging, and other traditional hurdle approaches to ensure integrity and shelf life.[4]

## Low-Temperature Storage

For mesophilic spore-forming species, temperatures below 15 °C are generally thought to prevent spores from germinating. This is why in a laboratory setting most spore crops are suspended in water and stored under refrigeration with the assumption that the spore crop concentration will remain stable until use.[4] Low temperatures and limited nutrients prevent germination; however, in the case of psychrotolerant spore-forming species, including *B. cereus*, temperatures at or above 6 °C[14] may allow for spore germination (albeit slowly) with outgrowth and perhaps permit cell multiplication in nutrient-rich environments. This possibility is why psychrotolerant spore-forming species are of concern.[4] Such would be the case with minimally processed foods not heated or significantly heat processed prior to eating.

## *BACILLUS CEREUS* AND ITS ENDOSPORES

*Bacillus cereus* is a motile Gram–positive spore-forming bacterium that is a well-established food-borne pathogen.[7,16,17] It is found throughout nature but is most commonly isolated from soil and plants.[3,18] Food-borne illnesses caused by *B. cereus* are directly related to the production of two toxin types: an emetic-type enterotoxin and a group of several diarrheagenic-type enterotoxins.[7,17] The enterotoxins cause various gastrointestinal illnesses including diarrhea and emesis.[19] The emetic-type toxin, also known as cereulide, is a thermostable cyclic peptide.[7,19] The enterotoxins responsible for the diarrheagenic symptoms caused by *B. cereus* are hemolysin (Hbl), nonhemolytic enterotoxin (Nhe), and cytotoxin (CytK).[19,20] The cell wall of vegetative *B. cereus* is also covered by proteins (called the S-layer) that play a role in cell adhesion and contribute to the virulence of the organism.[16]

Since the emetic toxin of *B. cereus* is heat stable, it can remain stable after cooking or heating. Cereulide production does not occur until the stationary growth phase, and therefore, high counts of vegetative cells or spores able to germinate in foods are required for cereulide intoxication to occur.[21]

Cereulide toxin is absorbed from the gut into the bloodstream and induces emetic-like symptoms including nausea and vomiting through stimulation of the vagus nerve.[22] Ingestion of approximately $\leq 8\,\mu g\,kg^{-1}$ body weight of cereulide toxin within a food product is required to cause illness in humans.[22] Cereulide toxin affects the mitochondria by acting as a potassium ion channel former,[7,19,23] causing apoptosis of human natural killer cells.[24] Mesophilic strains of *B. cereus* can produce cereulide only at temperatures above 10–15 °C.[21] This is why the toxin is mostly associated with foods that are improperly cooled and stored, such as rice and pasta.

The diarrheagenic toxin is heat labile and can be destroyed by heating or cooking. This toxin is produced during the exponential growth phase and can cause intoxication when present in raw or minimally processed foods that do not require heating.[25] Since spores of *B. cereus* are capable of surviving heat treatment and the acidic environment of the stomach, diarrhea–like symptoms can occur when spores of *B. cereus* are consumed in a raw or unprocessed food product and enter the small intestines, where spores can germinate and multiply, enabling the production of the diarrheagenic toxin. Depending on the amount of bacteria present in the food product, sometimes both sets of symptoms (emesis and diarrhea) can develop. This phenomenon is known as two–bucket disease, such as occurs with *Staphylococcus aureus* intoxication. In either case, symptoms from either *B. cereus* toxin should resolve within 24–48 h from onset.

The foods most frequently associated with *B. cereus* intoxication include milk, vegetables, rice, potatoes, grains, cereals (including batters, mixes, and breadings), spices, and various sauces.[26] *Bacillus cereus* is not nutritionally fastidious, which is why *B. cereus* can replicate in soil and low–nutrient foods including rice and pasta.[16] The reservoir for *B. cereus* is the soil, where transmission of the organism can occur through various vectors[27]; however, the most common vector is through food. As with many self–limiting food-borne gastrointestinal illnesses, people experiencing *B. cereus* intoxication usually do not seek medical attention because of the generally short duration of the illness and nonfatal symptoms. Even if medical attention is given, the illness is not reportable.[19] Lack of testing, reporting, and surveillance of the illness has led to underestimation of the actual incidence of food-borne illness caused by *B. cereus*.

## Bacterial Spores in the Food Industry

As noted earlier, spores possess extreme resistance to heat, pressure, extremes of pH, disinfectant chemicals, irradiation, desiccation, infectious agents, and just about any stress agent or conditions imaginable, including being able to survive over extremely long time periods. Spores themselves are generally of no concern unless they are able to germinate within the food or after consumption. Therefore, understanding the germination and outgrowth of spores is of fundamental importance.

Spores can survive for long periods of time in food products, particularly in foodstuffs in which nutrient content is low or nonexistent.[6] When these spores germinate, food-borne illness can occur.[5] It would be ideal to trigger germination of spores present in the food product prior to or during any preservation treatment, since spores are much less resistant and more susceptible to inactivation after they have germinated.[28] Although this strategy seems simple, germination rates vary, and a small percentage of spores commonly germinate extremely slowly or not at all after exposure to germinants.[8,28] Such spores are known as superdormants.

## Superdormant Spores

Until recently, studies on bacterial spores have primarily focused on populations and have neglected spores that either fail to germinate or germinate extremely slowly.[28] A simple method for isolation of superdormant spores was developed by Ghosh and Setlow[28] for spores of *Bacillus subtilis*, *Bacillus megaterium*, and *B. cereus*. This method, called buoyant density centrifugation, separates dormant spores from germinating spores and debris. After this type of centrifugation, dormant spores are in the pellet and germinated spores float.[28] Through multiple cycles of heat shock, germination, and buoyant density centrifugation, the majority of spores will have germinated, leaving spores that have either failed to germinate or take longer to germinate than the remaining spore population. Owing to the development of this method of isolation, there have been many discoveries surrounding the identification and characterization of superdormant spores (Table 1).

It appears that the physiological state for superdormancy is similar for all *Bacillus* species.[28] Recent studies provide evidence that suggest that one reason for superdormancy is a reduced level of germinant receptors.[8,28,29] Because superdormant spores are not genetically different from the remaining spore population, there is no current method to determine whether a spore is superdormant. Simply, a spore that fails to germinate or germinates much more slowly compared to the spore population from which it was isolated can be classified as superdormant.

It has been demonstrated that sublethal heat treatment prior to germination decreases the yield of superdormant spores; however, superdormant spores still show a higher temperature optimum for heat activation than the remainder of the spore population.[28] It also appears that superdormant

**Table 1** Characteristics that identify superdormant spores as well as the significance of each of these discoveries

| Characteristic | Significance |
|---|---|
| SD spores germinate poorly with single-nutrient germinants[28] | Superdormants lack nutrient GRs |
| SD spores germinate normally with combinations of nutrients[28] | More nutrients activate more nutrient receptors |
| SD spores germinate normally with Ca-DPA and dodecylamine[28] | These germinants do not trigger nutrient receptors |
| SD spores germinate normally with 150 MPa pressure[8] | Surprising—germination triggered through nutrient receptors |
| SD spores germinate normally with 500 MPa pressure[8] | Not surprising—germination not triggered through nutrient receptors |
| A 1-log reduction in SD spores is achieved with 120 ppm ozone followed by high-pressure processing[54] | This is significant considering that the number of SD spores within a product is probably low |
| Levels of GRs, levels of a small protein that may be a GR subunit, and a *gerD* mutation may affect SD spore germination by mHP[31] | Provides new information on factors that modulate high-pressure germination of SD spores |

SD, superdormant; GR, germinant receptor; Ca-DPA, calcium dipicolinic acid; mHP, moderate high pressure.

spores have a greater wet–heat resistance and lower core water contents.[30] Superdormant spores germinate poorly in the presence of nutrient germinants compared to other germinants, such as dodecylamine or calcium dipicolinic acid (Ca-DPA). This is not surprising since germination by dodecylamine or Ca-DPA does not require nutrient binding by receptors nor does it require prior heat activation.[8,28]

In agreement with the findings of Ghosh and Setlow,[28] Zheng et al.[29] found that a number of factors increase the rate of spore germination, including heat activation and an increased level of germinant receptors. It is suspected that different germinant receptors within an individual spore interact through aggregation that could potentially amplify signals from large numbers of germinant receptors.[28] The lack of these nutrient receptors may inhibit the amplification of this germination signal and may explain why higher yields of superdormant spores are observed with *Bacillus* strains that lack one or more germinant receptor.[28] To understand why spore germination is determined by the level of germinant receptors on an individual spore, it must first be determined how ligand binding to germinant receptors triggers spore germination.[28] It may be possible that a low level of GerD receptors on a spore can contribute to spore superdormancy, as reported by Ghosh and Setlow,[28] who found the rate of germination by nutrients increased in a spore population containing higher numbers of GerD receptors. It has also been proposed that a *gerD* gene mutation as well as the presence of a small protein that is a germinant receptor subunit may also play a role in spore superdormancy.[31] Heterogeneity in a spore population, resulting in varying rates of germination among individual spores, may also be due to adaptation of a particular bacterial species. Spores that germinate more slowly or at a reduced rate compared to the majority of the population are more likely to survive environmental changes by which the majority of germinating spores are inactivated, thus increasing the likelihood of survival for the entire population.[28]

Hydrostatic pressure inactivation studies on *B. cereus* and *B. subtilis* by Wei et al.,[8] using relatively low and high pressure magnitudes, demonstrated almost identical results for both species. Both germinated normally at 150 and 500 MPa.[8] It was not surprising that spores germinated after exposure to 500 MPa since spore inactivation at this pressure level does not affect nutrient germinant receptors and does not require heat activation[8]; however, it was surprising that superdormant spores germinated at 150 MPa since spore germination at this pressure magnitude requires activation of nutrient receptors.[8] Further studies on pressure inactivation of superdormant spores are essential for high-pressure processing (HPP) to be more efficiently utilized by the food industry. Superdormant spores are no doubt a significant contributing factor in the incomplete sterilization of low-acid foods using high-hydrostatic-pressure processing. Wei et al.[8] also found that superdormant spores of *Bacillus* species were able to germinate normally when exposed to peptidoglycan fragments and bryostatin. This was expected since germination initiation by these two agents appears to be triggered by eukaryotic-like membrane-bound serine/threonine protein kinase domains[32] and is not a result of interaction with germinant receptors.

## NONTHERMAL PROCESSING TECHNOLOGIES FOR *BACILLUS CEREUS* SPORES

### Ozone

Spores have been shown to be inactivated by several oxidizing agents, including chlorine dioxide, hydrogen peroxide, organic hydroperoxides, ozone, and sodium hypochlorite. Aqueous ozone has a higher potential than most oxidizing agents to inactivate spores.[33,34] Studies involving the inactivation of spores by oxidizing agents suggest that inactivation is a result of oxidative damage to the spore's inner membrane.

Young and Setlow[35] found that (1) when treated with ozone, spores of *B. subtilis* were more easily inactivated when they were uncoated prior to ozone treatment; (2) the spores did not germinate with nutrient germinants or Ca-DPA after ozone treatment; and (3) germination of the spores with ozone did not cause release of DPA from the spore's inner core.[35] The authors concluded that spores are not inactivated with ozone by DNA damage and that the major resistance factor of spores to ozone is the spore coat.[35]

Studies performed by Cortezzo et al.[36] confirmed that ozone causes damage to the spore's inner membrane because ozone-treated spores of *B. subtilis* were more easily penetrated by methylamine and germinated faster with dodecylamine. Since the inner membrane of the spore is known to be a barrier to methylamine, this study demonstrated that the inner membrane was damaged by its inability to prevent methylamine from leaking through the barrier into the spore. Damage to the spore's inner membrane via oxidization can have several effects, including inability to germinate, spore death after germination, and spore lysis.[36] More interestingly, the authors found that spore survivors of ozone treatment exhibited increased sensitivity to inactivation by a normally minimal heat treatment. Spores treated with ozone were also more sensitive to NaCl in plating media than untreated spores. Since heat treatment and NaCl treatment are not lethal to spores under normal conditions, these findings further confirm that ozone causes damage to the spore's inner membrane, making the spores more sensitive to these treatments. The authors hypothesize that ozone treatment causes damage to key proteins in the spore's inner membrane, although more research needs to be done to determine what these proteins are.[36]

## High-Pressure Processing

The use of pressure in food processing was first used in 1899 by Hite, who found that pressure treatment of milk could increase its shelf life.[37] Over a century later, application of pressure by the food-processing industry to extend product shelf life and safety now occurs. With consumer demands for fresh food products on the rise, interest in nonthermal processing techniques that will not damage the sensory qualities of food products while reducing microbial contamination continues. High hydrostatic pressure is currently one of the nonthermal processing methods utilized by the food industry.

High-pressure-processed foods were first introduced to the Japanese market in 1990; now, pressure-treated foods are available around the world.[37] Currently HPP is used to commercially process guacamole, presliced deli meats, juices, and oysters. Researchers and producers are interested in expanding the application of HPP to a wider variety of foods.

### Vegetative Cells

Vegetative cells are inactivated by HPP through a variety of mechanisms, most of which involve the cell membrane. A pressure-induced decrease in cell volume can permeabilize the cell membrane and lead to death.[37] Inactivation of bacteria by HPP is dependent on the manner in which the pressure is applied (cyclic or continuous), treatment temperature, treatment time, strain, cell shape, Gram stain type, growth stage, and treatment medium.[37] HPP of vegetative bacterial cells is generally more effective at higher temperatures unless the bacterial species contains heat-shock proteins, in which case heat would cause a baroresistant effect.[38] HPP applied in cyclic phases rather than continuously also tends to be somewhat more effective at inactivating bacterial species. Rod-shaped cells are more sensitive to HPP than cocci.[39] The presence of various ions in the medium may or may not induce baroresistance or sensitivity, depending on the microorganism. In the presence of low water activity and large amounts of sorbitol and glycerol, a baroprotective effect on the inactivation of microorganisms can take place.[40] HPP treatment has also been hypothesized to cause cleavage of the cell's DNA, leading to cell death.[41] The resistance of microorganisms to HPP is largely dependent on the species and strain of the microorganism and is extremely variable,[42] but most vegetative cells of bacteria and yeast show inactivation at pressures around 300–400 MPa at ambient temperature.[43]

### Spores

One of the current disadvantages of HPP is its inability to inactivate spores by pressure alone without altering the sensory qualities of the product.[37,44] Complete inactivation of spores remains a key necessity to produce shelf-stable low-acid pressure-treated foods. Thus, it is important to understand the physiology of spores, especially pertaining to spore inactivation by HPP.[44]

It is currently hypothesized that spore inactivation via HPP is caused by temperature- and pressure-induced spore germination in which spores lose their resistance and are inactivated owing to increased sensitivity to pressure.[44]

This process is more specifically known as electrostriction, by which a pressure-induced decrease in water volume of the cell causes a local collapse of the bulk water structure of the cell, thus beginning the germination process.[37] While spores can be resistant to pressures as high as 1200 MPa, low-pressure treatments from 50 to 300 MPa can induce spore germination at higher temperatures.[37,44] Lower pressures can trigger spore germination through activation of the nutrient receptors on the inner membrane of dormant[45–47] and superdormant spores.[8] Small acid-soluble protein degradation, which normally accompanies nutrient-induced germination, has also been observed to take place in spores treated with moderate pressures but not high pressures.[48] Extremely high pressures trigger spores, dormant or superdormant, to germinate, causing release of Ca-DPA from the spore core.[44] Spores germinated with high pressures are able to complete germination but go through outgrowth much more slowly than spores treated at lower pressures.[48]

There are several other factors that can affect the germination of spores with HPP. As with vegetative cells, spore germination and inactivation with HPP are also more effective when applied in a cyclic fashion.[49,50] Sporulation temperature has also been demonstrated to influence the HPP inactivation of spores of *B. cereus*.[42] In a study by Raso et al.,[42] spores that were initially sporulated at 37 °C were more significantly germinated and inactivated with HPP compared to spores that were initially sporulated at 20 °C. Wuytack et al.[48] demonstrated that *B. subtilis* spores exposed to HPP at >200 MPa were more sensitive to pressure, UV light, and hydrogen peroxide compared to spores not pretreated with pressure.

### Superdormant Spores

Since spore germination with moderate pressures is triggered by activation of nutrient receptors, spores with an increased number of germinant receptors will be more easily inactivated by moderate pressures.[46–48,51] Therefore; it would be expected that superdormant spores, which have a decreased number of germinant receptors, would germinate poorly with low or moderate pressures.

In a study by Wei et al.,[8] superdormant spores of *B. subtilis* and *B. cereus* that were not exposed to heat activation germinated normally, or similar to the rest of the spore population, at a pressure of 500 MPa. According to these authors, this result was expected since spore germination at this level of pressure does not require activation of nutrient germinant receptors. More interestingly, superdormant spores germinated normally at 150 MPa, which was surprising to the authors because spore germination induced at this level of pressure requires activation of the nutrient germinant receptors, and spore germination induced by this level of pressure is increased by prior heat activation.[8] The findings of this study were significant in that the factors responsible for spore germination by pressure were not identical to those responsible for spore superdormancy or nutrient germination.[8] It is an important finding for the food industry that superdormant spores are not the cause of incomplete germination of *Bacillus* species by HPP. More recently, it was discovered that the level of germinant receptors on the spore's inner membrane, levels of a small protein that may be a germinant receptor subunit, and a *gerD* mutation may affect superdormant spore germination by moderate high pressure.[31] It is important to determine the cause of incomplete germination and inactivation of spores by HPP so that this nonthermal processing technology can be more widely utilized by the food industry.

## USE OF HURDLE TECHNOLOGY TO REDUCE THE NUMBER OF SPORES IN FOODS

Hurdle technology involves the combination of processing technologies to establish hurdles for microbial growth and/or survival.[52] There are some hurdles, or technologies, that are considered high hurdles and some that are considered low.[53] Owing to synergistic effects of treatments, a combination of low hurdles may be as successful as the application of a single high hurdle.[53] The use of combined milder processing techniques may not only challenge the survival of bacterial spores within the product, similar to more intense individual processing techniques, but may also help preserve the sensory qualities of the food product. However, low hurdles might not be sufficient for making nutrient-rich foods safe.[53] A study by Markland et al.[54] demonstrated that a combined hurdle technology of ozone and HPP could reduce the number of spores in a superdormant spore population by up to 1log, which may be significant considering that the amount of superdormant spores within a food product at any particular time is probably low.

## CONCLUSIONS

Bacterial spore formers, specifically *B. cereus*, and their means of inactivation continue to serve as a major challenge for the food industry. Although spores can be inactivated by cooking, heat can often destroy the organoleptic properties of certain foods such as raw vegetables. HPP and other nonthermal technologies have shown promise for reducing the number of spores within food products; however, there often remains a specific population of spores within the product that germinate more slowly, if at all, compared to the rest of the spore population. With an increase in the consumption of minimally processed foods that are not heated prior to consumption, there needs to be more research to determine the means by which bacterial spores can be inactivated within these products without deteriorating the organoleptic quality of that product. The development of specific hurdle technology strategies may also help food processors overcome the obstacle of complete spore inactivation, including superdormant spores.

## REFERENCES

1. Nissen H, Rosnes JT, Brendehaug J, Kleiberg GH. Safety evaluation of sous vide-processed ready meals. *Lett Appl Microbiol* 2002;**35**(5):433–8.
2. Samapundo S, Everaert H, Wandutu J, Rajkovic A, Uyttendaele M, Devileghere F. The influence of headspace and dissolved oxygen level on growth an haemolytic BL enterotoxin production of a psychrotolerant *Bacillus weihenstephanensis* isolate on potato based ready to eat foods. *Food Microbiol* 2010:1–7. [Reprint].
3. Valero M, Fernández PS, Salmerón MC. Influence of pH and temperature on growth of *Bacillus cereus* in vegetable substrates. *Int J Food Microbiol* 2003;**82**(1):71–9.
4. Markland SM, Farkas DF, Kniel KE, Hoover DG. Pathogenic psychrotolerant spore-formers: an emerging challenge for low-temperature storage of minimally processed foods. *Foodborne Pathog Dis* 2013;**10**(5):413–9.
5. Setlow P. Spore germination. *Curr Opin Microbiol* 2003;**6**:550–6.
6. Coleman WH, Chen D, Li Y, Cowan A, Setlow P. How moist heat kills spores of *Bacillus subtilis*. *J Bacteriol* 2007;**189**(23):8458–66.
7. De Vries YP, Van Der Voort M, Wijman J, Van Schaik W, Hornstra LM, De Vos WM, et al. Progress in food-related research focusing on *Bacillus cereus*. *Microbes Environ* 2004;**19**(4): 265–9.
8. Wei J, Shah IM, Ghosh S, Dworkin J, Hoover DG, Setlow P. Superdormant spores of *Bacillus* species germinate normally with high pressure, peptidoglycan fragments, and bryostatin. *J Bacteriol* 2010;**192**:1455–8.
9. Carlin F, Albagnac C, Rida A, Guinebretiere MH, Couvert O, Nguyen-The C. Variation of cardinal growth parameters and growth limits according to phylogenetic affiliation in the *Bacillus cereus* group. Consequences for risk assessment. *Food Microbiol* 2013;**33**(1): 69–76.
10. Schraft H, Steele M, McNab B, Odumeru J, Griffiths MW. Epidemiological typing of *Bacillus* spp. isolated from food. *Appl Environ Microbiol* 1996;**62**(11):4229–32.
11. Daffonchio D, Cherif A, Borin S. Homoduplex and heteroduplex polymorphisms of the amplified ribosomal 16S-23S internal transcribed spacers describe genetic relationships in the "*Bacillus cereus* group". *Appl Environ Microbiol* 2000;**66**(12):5460–8.
12. Sorokin A, Candelon B, Guilloux K, Galleron N, Wackerow-Kouzova N, Ehrlich SD, et al. Multiple-locus sequence typing analysis of *Bacillus cereus* and *Bacillus thuringiensis* reveals separate clustering and a distinct population structure of psychrotrophic strains. *Appl Environ Microbiol* 2006;**72**(2):1569–78.

13. Guinebretiere MH, Thompson FL, Sorokin A, Normand P, Dawyndt P, Ehling-Schulz M, et al. Ecological diversification in the *Bacillus cereus* Group. *Environ Microbiol* 2008;**10**(4): 851–65.

14. Lechner S, Mayr R, Francis KP, Prub BM, Kaplan T, Weibner-Gunkel E, et al. *Bacillus weihenstephanensis* sp. nov. is a new psychrotolerant species of the *Bacillus cereus* group. *Int J Sytem Bacteriol* 1998;**48**:922–31.

15. Baron F, Cochet MF, Grosset N, Madec MN, Briandet R, Dessaigne S, et al. Isolation and characterization of a psychrotolerant toxin producer, *Bacillus weihenstephanensis*, in liquid egg products. *J Food Prot* 2007;**70**(12):2782–91.

16. Kotiranta A, Lounatmaa K, Haaplasalo M. Epidemiology and pathogenesis of *Bacillus cereus* infections. *Microbes Inf* 2000;**2**:189–98.

17. Chorin E, Thuault D, Cléret JJ, Bourgeois CM. Modelling *Bacillus cereus* growth. *Int J Food Microbiol* 1997;**38**(2–3):229–34.

18. Priest FG, Barker M, Baillie LWJ, Holmes EC, Maiden MCJ. Population structure and evolution of the *Bacillus cereus* group. *J Bacteriol* 2004;**186**(23):7959–70.

19. Lucking G, Dommel MK, Scherer S, Fouet A, Ehling-Schulz M. Cereulide synthesis in emetic *Bacillus cereus* is controlled by the transition state regulator AbrB, but not by the virulence regulator PlcR. *J Microbiol* 2009;**155**:922–31.

20. Ehling-Schulz M, Svensson M, Guinebretiere MH, Lindback T, Andersson M, Schulz A, et al. Emetic toxin formation of *Bacillus cereus* is restricted to a single evolutionary lineage of closely related strains. *J Microbiol* 2005;**151**:183–97.

21. Thorsen L, Budde BB, Henrichsen L, Martinussen T, Jakobsen M. Cereulide formation by *Bacillus weihenstephanensis* and mesophilic emetic *Bacillus cereus* at temperature abuse depends on pre-incubation conditions. *Int J Food Microbiol* 2009;**134**(1–2):133–9.

22. Jaaskelainen EL, Teplova V, Andersson MA, Andersson LC, Tammela P, Andersson MC, et al. In vitro assay for human toxicity of cereulide, the emetic mitochondrial toxin produced by food poisoning *Bacillus cereus*. *Toxicol Vitro* 2003;**17**(5–6):737–44.

23. Mikkola R, Saris NE, Grigoriev PA, Andersson MA, Salkinoja-Salonen MS. Ionophoretic properties and mitochondrial effects of cereulide: the emetic toxin of *B. cereus*. *Eur J Biochem* 1999;**263**(1):112–7.

24. Paananen A, Mikkola R, Sareneva T, Matikainen S, Hess M, Andersson M, et al. Inhibition of human natural killer cell activity by cereulide, an emetic toxin from *Bacillus cereus*. *Clin Exp Immunol* 2002;**129**(3):420–8.

25. Fermanian C, Fremy JM, Claisse M. Effect of temperature on the vegetative growth of type and field strains of *Bacillus cereus*. *Lett Appl Microbiol* 1994:16.

26. Doona CJ, Feehery FE. Inactivation of *Bacillus cereus* by high hydrostatic pressure. In: Al-Holy MA, Lin M, Rasco BA, editors. *High pressure processing of foods*. Ames: Blackwell; 2007.

27. Abee T, Groot MN, Tempelaars M, Zwietering M, Moezelaar R, van der Voort M. Germination and outgrowth of spores of *Bacillus cereus* group members: diversity and role of germinant receptors. *Food Microbiol* 2011;**28**(2):199–208.

28. Ghosh S, Setlow P. Isolation and characterization of superdormant spores of *Bacillus* species. *J Bacteriol* 2009;**191**(6):1787–97.

29. Zhang P, Garner W, Yi X, Yu J, Li YQ, Setlow P. Factors affecting variability in time between addition of nutrient germinants and rapid dipicolinic acid release during germination of spores of *Bacillus* species. *J Bacteriol* 2010;**192**(14):3608–19.

30. Ghosh S, Zhang P, Li YQ, Setlow P. Superdormant spores of *Bacillus* species have elevated wet-heat resistance and temperature requirements for heat activation. *J Bacteriol* 2009;**191**(18):5584–91.

31. Doona CJ, Ghosh S, Feeherry FF, Ramirez-Peralta A, Huang Y, Chen H, et al. High pressure germination of *Bacillus subtilis* spores with alterations in levels and types of germination proteins. *J Appl Microbiol* 2014;**117**(3):711–20.

32. Shah IM, Laaberki MH, Popham DL, Dworkin J. A eukaryotic-like Ser/Thr kinase signals bacteria to exit dormancy in response to peptidoglycan fragments. *Cell* 2008;**135**(3): 486–96.

33. Menzel DB. Oxidation of biologically active reducing substances by ozone. *Arch Environ Health* 1971;**23**(2):149–53.

34. Kim JG, Yousef AE, Khadre MA. Ozone and its current and future application in the food industry. *Adv Food Nutr Res* 2003;**45**:167–218.

35. Young SB, Setlow P. Mechanisms of *Bacillus subtilis* spore resistance to and killing by aqueous ozone. *J Appl Microbiol* 2004;**96**(5):1365–2672.

36. Cortezzo DE, Koziol-Dube K, Setlow B, Setlow P. Treatment with oxidizing agents damages the inner membrane of spores of *Bacillus subtilis* and sensitizes spores to subsequent stress. *J Appl Microbiol* 2004;**97**(4):838–52.

37. San Martin MF, Barbosa-Canovas GV, Swanson BG. Food processing by high hydrostatic pressure. *Crit Rev Food Sci Nut* 2010;**46**(6):627–45.

38. Iwahashi H, Obuchi K, Fuji S, Fujita K, Komatsu Y. The reason why trehalose is more important for barotolerance than hasp104 in *Saccharomyces cerevisiae*. In: Heremans K, editor. *High pressure research in the biosciences and biotechnology*. Belgium: Leuven University Press; 1996.

39. Ludwig H, Schreck C. The inactivation of vegetative bacteria by pressure. In: Heremans K, editor. *High pressure research in the biosciences and biotechnology*. Belgium: Leuven University Press; 1996.

40. Hayert M, Perrier-Cornet JM, Gervais P. Why do yeasts die under pressure? In: Heremans K, editor. *High pressure research in the biosciences and biotechnology*. Belgium: Leuven University Press; 1996.

41. Chilton P, Isaacs NS, Mackey B, Stenning R. The effects of high hydrostatic pressure on bacteria. In: Heremans K, editor. *High pressure research in the biosciences and biotechnology*. Belgium: Leuven University Press; 1996.

42. Raso J, Gongora-Neito MM, Barbosa-Canovas GV, Swanson BG. Influence of several environmental factors on the initiation of germination and inactivation of *Bacillus cereus* by high hydrostatic pressure. *Int J Food Microbiol* 1998;**44**(1–2):125–32.

43. Knorr D. High pressure effects on plant derived foods. In: Ledward DA, Johnston DE, Earnshaw RG, Hasting M, editors. *High pressure processing of foods*. Nottingham: Nottingham University Press; 1995.

44. Black EP, Setlow P, Hocking AD, Stewart CM, Kelly AL, Hoover DG. Response of spores to high-pressure processing. *Comp Revs Food Sci Food Saf* 2007;**6**:103–19.

45. Wuytack EY, Soons J, Poschet F, Michiels CW. Comparative study of pressure- and nutrient-induced germination of *Bacillus subtilis* spores. *Appl Environ Microbiol* 2000;**66**(1):257–61.

46. Paidhungat M, Setlow B, Daniels WB, Hoover D, Papafragkou E, Setlow P. Mechanisms of induction of germination of *Bacillus subtilis* spores by high pressure. *Appl Environ Microbiol* 2002;**68**(6):3172–5.

47. Black EP, Koziol-Dube K, Guan D, Wei J, Setlow B, Cortezzo DE, et al. Factors influencing germination of *Bacillus subtilis* spores via activation of nutrient receptors by high pressure. *Appl Environ Microbiol* 2005;**71**(10):5879–87.

48. Wuytack EY, Boven S, Michiels CW. Comparative study of pressure-induced germination of *Bacillus subtilis* spores at low and high pressures. *Appl Environ Microbiol* 1998;**64**(9):3220–4.

49. Hayakawa I, Kanno T, Yoshiama K, Fujio Y. Oscillatory compared with continuous high pressure sterilization on *Bacillus stearothermophilus* spores. *J Food Sci* 2006;**59**(1):164–7.

50. Palou E, Lopez-Malo A, Barbosa-Canovas GV, Welti-Chanes J, Davidson PM, Swanson BG. Effect of oscillatory high hydrostatic pressure treatments on *Byssochlamys nivea* ascospores suspended in fruit juice concentrates. *Lett Appl Microbiol* 1998;**27**(6):375–8.

51. Pelczar PL, Igarashi T, Setlow B, Setlow P. Role of GerD in germination of *Bacillus subtilis* spores. *J Bacteriol* 2007;**189**(3):1090–8.

52. Leistner L. Principles and applications of hurdle technology. In: Gould GW, editor. *New methods for food preservation*. London: Blackie Academic and Professional; 1995. p. 1–21.

53. Leistner L, Gorris LGM. Food preservation by hurdle technology. *Trends Food Sci Tech* 1995;**6**:41–6.

54. Markland SM, Kniel KE, Setlow P, Hoover DG. Nonthermal inactivation of heterogeneous and superdormant spore populations of *Bacillus cereus* using ozone and high pressure processing. *IFSET* 2013;**19**(44–49).

# CHAPTER 5

# *Bacillus cereus* Food-Borne Disease

**Roberta Marrollo**
Clinical Microbiology and Virology, Laboratory of Bacteriology and Mycology, Civic Hospital of Pescara, Pescara, Italy

## SUMMARY

*Bacillus cereus* is ubiquitous in nature and is found on soil, on plants, and in the enteric tract of insects and mammals. From these niches it is easily spread to food products, causing an emetic or diarrheal syndrome after ingestion. The former is due to cereulide, a small toxin whose genetic determinants are plasmid borne. The diarrheal syndrome is instead caused by vegetative cells, ingested in the form of spores of viable cells, that are thought to produce protein enterotoxins in the small gut. Pathogenesis of the diarrheal disease relies on three pore-forming cytotoxins, which are nonhemolytic enterotoxin (Nhe), hemolysin BL (Hbl), and cytotoxin K. Nhe and Hbl are homologous three-component toxins related to the toxin cytolysin A of *Escherichia coli*.

## BACKGROUND

*Bacillus cereus* is frequently found in food production environments owing to its highly adhesive endospores that can spread to all kinds of foods. The organism produces a range of pathogenic factors that may cause food-borne diseases in humans and is one of the major food-related pathogenic bacteria in general; the enteric disease is, however, in most cases mild and of short duration.[1–4] It may occur in the form of a diarrheal infection or an emetic syndrome, which develops by virtue of diverse types of toxins. The emetic toxin, which causes vomiting, is a small ring-form peptide[5]; the diarrheal syndrome, instead, is due to one or more protein enterotoxins, which are thought to elicit diarrhea by disrupting the integrity of the plasma membrane of small gut epithelial cells. The three pore-forming toxins that are known to be involved in the diarrheal infection are the cytotoxins hemolysin BL (Hbl), nonhemolytic enterotoxin (Nhe), and cytotoxin K (CytK).[3,6,7]

*The Diverse Faces of Bacillus cereus*
ISBN 978-0-12-801474-5
http://dx.doi.org/10.1016/B978-0-12-801474-5.00005-0

## FOOD CONTAMINATION

*Bacillus cereus* is isolated from a wide range of foods along with food ingredients, such as rice, dried foods, vegetables, spices, milk, and dairy products.[8] Cross-contamination can therefore distribute cells or spores to other foods.[9–11] At the time of harvest, cells or spores are known to potentially accompany plant material into food production environments and then establish on food-processing equipment. In dairy products, *B. cereus* can cause a defect known as sweet curdling. Moreover, spores or cells may colonize udders of cows during grazing[12] or enter the dairy farm via bedding material or feed.[8]

The spores produced by *B. cereus* are a huge advantage for this bacterium, as they permit attachment to surfaces and survival upon heat treatment or other procedures that are aimed at removing vegetative microorganisms, which could otherwise outgrow *B. cereus*.[13–15] *Bacillus cereus* spores are not necessarily removed by regular surface cleaning[12,16]; also, the ability of this pathogen to enter yet another lifestyle, thus forming biofilm communities,[17] is likely to be important for persistence in food industry equipment (e.g., dairy pipelines); in fact, the mentioned biofilm prevents inactivation by sanitizers of both spores and vegetative cells.[18]

Today's large-scale technology of food production, which extends refrigeration as a means of conservation, has led to the selection of a cold niche well suited to bacteria that are not very competitive but can, however, survive heat and grow at low temperatures. This is the case of *Bacillus weihenstephanensis* along with *B. cereus* and other *Bacillus* species, frequently isolated from dairy products and from those environments in which cooling is extensively used as a means of controlling bacterial growth. Additionally, lightly heat-treated foods characterized by extended refrigerated storage represent a novel, favorable environment for species of the *B. cereus* group.[19–21]

If we consider the almost ubiquitous distribution of *B. cereus* in the natural environment and the resilient spores it produces, as well as the nonfastidious nature of this microorganism, it emerges clearly that each type of food with pH >4.8 may be a potential vehicle or harbor food spoilage, thus supporting food-borne infection.[10] Failure by consumers to carry out correct food preparation procedures (for example, inadequate or slow cooling, storage at ambient temperature, or sitting at <60 °C for a prolonged time) can be responsible for *B. cereus* replication and be behind cases of food-borne disease by this organism.[10]

## FOOD-BORNE INFECTION

*Bacillus cereus* may cause two diverse and distinct types of food-borne disease, that is, an emetic and a diarrheal syndrome. Both are usually mild and self-limiting, although more severe and even fatal cases have been reported.[2–4,22] *Bacillus cereus* was acknowledged as a cause of food poisoning in the 1950s, the first described diarrheal outbreaks being observed in Norwegian hospitals in 1947–1949.[23,24] No specific populations of patients have been described as being particularly predisposed to *B. cereus* food-borne disease; nonetheless, lowered stomach acidity (such as in the case of elderly people or those suffering from achlorhydria) may reasonably represent a risk factor.[25]

Toxins on which *B. cereus* food poisoning relies are described in a separate chapter of this volume. However, it is of interest to emphasize that the emetic syndrome was first recognized after the occurrence in the early 1970s of numerous outbreaks caused by eating cooked rice in the United Kingdom.[26] This disease is an intoxication due to *B. cereus* emetic toxin, which is named cereulide and is produced in foods before ingestion. The course of the syndrome is characteristic, as it presents with nausea and emesis only a few hours after the meal. The incubation time, nowadays, is reported to be 0.5–6 h, and the clinical picture is currently considered to last usually 6–24 h.[5]

*Staphylococcus aureus* enterotoxins cause similar symptoms; however, in this case, emesis is generally accompanied by diarrhea.[27] Also, unfortunately, serious and even lethal cases of emetic *B. cereus* disease have been described more than once in the literature.[2,4,5,28,29]

The diarrheal disease is thought to represent a toxic infection caused by *B. cereus* vegetative cells that, after ingestion as spores or viable cells, produce protein enterotoxins in the small gut.[25,30,31]

The disease typically presents with watery diarrhea, abdominal pain, and sometimes nausea and emesis, and it is easy to confuse it with the food-borne syndrome due to the sporulating anaerobe *Clostridium perfringens* type A.[32] The incubation time is 8–16 h, 12 h on average, although in rare cases a longer incubation period has been observed. The disease duration is usually 12–24 h, but episodes lasting several days have been described as well.[2]

Many kinds of food may be the vehicle for *B. cereus*, such as rice, pasta, spices, meats (including poultry), and sprouts.[8–10,33] Also, different foods are more commonly associated with either of the two types of syndrome: in fact, the emetic type has frequently been related to fried and cooked rice, as well as pasta, pastry, and noodles,[10,11,34] while the diarrheal type is usually associated with proteinaceous foods, meat and milk products, along with sauces and vegetables, soups, and puddings.[8,10,11,33] Paradoxically, the emetic *B. cereus* strains more often contaminate starch-rich foods, although they are generally not able to hydrolyze starch.[5]

Different distributions between countries may be observed for the emetic and the diarrheal pictures caused by *B. cereus*, which probably reflects the different food vehicles the organism may use: in fact, in the United Kingdom and Japan the emetic syndrome is more prevalent,[10,35] whereas in North America and northern Europe the diarrheal form seems to dominate.[33]

Doses as low as $10^3$ *B. cereus* CFU (colony-forming units) $g^{-1}$ of contaminated food have been found in diarrheal disease cases, although lower quantities of spores compared with vegetative cells may probably be responsible for this syndrome, as spores can better tolerate the gastric acid.[10,25]

Instead, the amount of cells required to cause the emetic disease has not been defined yet, but in incriminated foods $10^3$–$10^{10}$ CFU $g^{-1}$ seems to be the currently accepted infective dose (in most episodes at least $10^5$ CFU $g^{-1}$).[10]

*Bacillus cereus* is an important agent of food-borne disease worldwide, although it may be greatly underreported as an enteric pathogen.[11,36] Several factors contribute to such an underreporting, including the generally short and mild course of the illness, which usually does not lead the patient to look for medical attention. Furthermore, both cases and outbreaks may sometimes not be attributed to *B. cereus*, since the clinical picture of the emetic disease can be easily confused with that caused by *S. aureus* intoxication, whereas the diarrheal syndrome shares symptoms with *C. perfringens* type A food poisoning, as mentioned above.[10,33] However, unfortunately, as the surveillance systems for food-borne disease differ between countries, it is hard to compare data and reach reliable incidence estimates.

## TOXINS

Although toxins responsible for *B. cereus* food-borne disease are discussed in a separate chapter, we would like to emphasize here, too, that the onset of the emetic disease caused by *B. cereus* is usually fast, starting generally from 0.5 to 6 h after meal ingestion, which indicates it is an intoxication due to the presence of a toxin preformed in the food. Cereulide is the emetic toxin and is encoded by the 24-kb cereulide synthetase (*ces*) gene cluster located on a megaplasmid related to pXO1. The plasmid was initially named pBCE4810 or pCER270. As cereulide is resistant to acid conditions, as well as proteolysis and heat, it tolerates gastric acid, gut proteolytic enzymes, and reheating of foods that have been stored at room temperature after the first heating.[37]

The mechanism behind cereulide activity is not yet completely clarified, but the toxin behaves as a cation ionophore, therefore it inhibits mitochondrial activity by inhibiting fatty acid oxidation, which is the reason for liver failure that occurred in two lethal cases of emetic disease. Accordingly, experimental mice that were injected intraperitoneally with a high dose of synthetic cereulide developed massive hepatocyte degeneration. The toxin has also been observed to inhibit human natural killer cells, with related subsequent immune system impairment.[37]

Cereulide production starts when the logarithmic phase ends, during the vegetative growth of *B. cereus*, with the highest level of production being observed at the early stationary growth phase; also, production is independent of sporulation. Synthesis occurs at 12–37 °C, although maximal production takes place between 12 and 22 °C. Two *B. weihenstephanensis* isolates, however, were found to produce cereulide at 8 °C.[37]

Different foods show variable ability to sustain production of the emetic toxin. In infant formulas, cereulide production was found to be influenced by the composition of the product, with a combination of cereal and dairy ingredients supporting higher levels of toxin production than nondairy and rice ingredients. Further studies documented only low cereulide levels in egg and meat products as well as in liquid foods, including milk and soy milk. Conversely, boiled rice and farinaceous food products were found to support high-level production of cereulide.[37]

Various levels of emetic toxin production have been shown between diverse strains and are maybe due to differences in production regulation; the *ces* genes show in fact only a low level of heterogeneity. Even environmental factors, including temperature, oxygen, and pH value, as well as the presence of certain amino acids, have been observed to influence cereulide production.[37]

The diarrheal syndrome instead was early on attributed to an enterotoxin, since culture filtrates of *B. cereus* were observed to be responsible for fluid accumulation in rabbit ileal loops. However, as *B. cereus* may secrete several cytotoxins and enzymes that may support the diarrheal disease, the identity of these compounds is still a controversial issue. The three cytotoxins Hbl, Nhe, and CytK are currently considered to be those on which *B. cereus* diarrheal disease relies. Hbl and Nhe are so-called three-component toxins, whereas the single-component CytK is part of the family of β-barrel pore-forming toxins. Additionally, numerous other cytotoxins, as well as hemolysins and degradative enzymes, have been detected that may perhaps contribute to the development of the diarrheal disease (i.e., cereo-lysin O, hemolysin II, hemolysin III, InhA2, and three phospholipases C).[37] The most relevant toxin among Hbl, Nhe, and CytK in the illness pathoge-nicity seems to vary between strains and, most likely, multiple toxins could act synergistically to cause disease.[37]

The cytotoxin CytK and all six components of Hbl and Nhe contain secretory signal peptides, meaning that they are secreted by the general secretory (Sec) pathway.[38] However, it has been proposed that the three Hbl components are secreted using the flagellar export apparatus.[39] This conclu-sion was supported by the observation of the absence of or reduced Hbl protein levels in supernatant from culture of certain nonflagellated mutant strains, as well as in an *flhF* mutant strain that also showed reduced numbers of flagella and altered motility.[40,41] However, these studies did focus on the issue of whether those reduced Hbl levels were mediated at the transcrip-tional, translational, or posttranslational level or were related to a secretion defect. Notably, a 2005 work revealed that the nonflagellated *flhA* mutant was characterized by a 50% reduction in *hbl* transcription.[42]

In 2008, it was documented, however, that all three Hbl proteins, together with the Nhe components and CytK, are actually secreted by the Sec path-way, as inhibition by azide of SecA (which is an essential component of the Sec translocase) led to a reduced secretion and subsequent intracellular accumulation of the toxin components.[37] Conversely, it was shown that the nonflagellated *flhA* strain showed decreased secretion of Hbl, Nhe, and CytK, but the absence of intracellular accumulation supported the fact that the lack of secreted toxin proteins was due to reduced production, not to secretion impairment.[37]

*Bacillus cereus* is not the only toxigenic species within the *B. cereus* group. In particular, several strains of *B. weihenstephanensis*, which was suggested to be a new species of the *B. cereus* group in 1998, have been found by Stenfors and colleagues to harbor the *nheBC* genes (like numerous cytotoxic *Bacillus thuringiensis* isolates) and, to a seemingly lower extent, *cytK*; also, production of two of the three components of each of the hemolytic and nonhemolytic enterotoxins was documented.[43,44] Finally, certain *B. weihenstephanensis* strains were observed, elsewhere in the literature, to possess the *hblA* gene.[43]

It should, however, be emphasized that positive polymerase chain reaction (PCR) results do not represent confirmation of the presence of a complete and functional gene, but they certainly provide a reliable indication as to whether the genetic element is at all present.[43] When PCR results synergize with other detection methodologies (i.e., antibody-based detection of enterotoxin proteins and the cytotoxicity assay), they nevertheless indicate that many *B. weihenstephanensis* strains indeed possess the genetic makeup for producing the mentioned virulence factors.[43,44]

Despite *Bacillus* species other than *B. cereus* being potentially toxigenic (i.e., the above-mentioned *B. thuringiensis* and *B. weihenstephanensis*), they are mostly considered to be harmless to humans, while *B. cereus* remains thus far the major cause of food-borne disease within the genus and, in general, is among the main bacterial agents of enteritis in humans.[37,45–78]

## DIFFERENTIAL DIAGNOSTICS

As mentioned above, *B. cereus* intoxications may resemble those by *S. aureus* and *C. perfringens*; 9.4 million food-borne disease cases per year (caused by a known organism) are estimated to occur annually in the United States; of these, 1.3 million (14%) are due to *B. cereus*, *C. perfringens*, or *S. aureus*.[79] All three of these pathogens cause enteric infections by virtue of preformed toxins produced in improperly handled food products (*B. cereus* and *S. aureus*) or in vivo inside the gut after ingestion of contaminated foods (*C. perfringens* and, hypothesized, diarrheal syndrome–related *B. cereus*).[79] Costs are estimated to range from US $166 per *B. cereus* illness to US $539 per *S. aureus* illness, with annual costs of diseases caused by these pathogens being estimated at US $523 million.[79]

Interestingly, symptoms, incubation period, illness duration, and suspected contaminated food products often overlap for these pathogens, so defining the etiology of a single episode or of a food-borne outbreak is usually challenging without laboratory confirmation either in clinical materials or in food samples.[79] Laboratory confirmation, additionally, may be complicated by the fact that patients often do not come to medical attention, stool specimens are not collected or are not examined for these pathogens or related enterotoxins, and implicated foods, finally, may no longer be available when the outbreaks are investigated.[79]

Some aspects may help in discriminating the poisoning etiology, presumptively. Vomiting seems to be the most useful sign that may discriminate *B. cereus* and *S. aureus* disease from *C. perfringens* illnesses; vomiting is in fact infrequent in *C. perfringens* outbreaks; *S. aureus* is instead commonly characterized by the shortest median incubation periods (<5 h).[79–85] Again, rice ingestion is almost always associated with *B. cereus* poisoning; meat and poultry may be involved, as well, but are more commonly associated with *C. perfringens* and *S. aureus* outbreaks.[79–85]

In *S. aureus* disease, pork, specifically ham, is frequently involved, along with poultry, egg and bakery products, casseroles, and milk and dairy products in general, although foods implicated in *S. aureus* outbreaks may differ between countries owing to diverse food consumption patterns.[79–85]

The *B. cereus* diarrheal syndrome is generally associated with meat products, along with sauces and vegetables, but also puddings and milk derivatives.[79–85]

Food contamination can occur at the source of food production or during processing, preparation, transport, and finally storage. In the event of *B. cereus* and *C. perfringens* infection contamination may occur in the environment, as both organisms are ubiquitous in soil, gut of animals, and several foods and ingredients.[79–85] Moreover, spores of *B. cereus* and *C. perfringens* are resistant to cooking, and, owing to improper preparation procedures, these pathogens can replicate, form heat-stable toxins, and withstand further cooking.[79–85]

Raw meat and dairy products can undergo contamination by virtue of staphylococcal carriage or infection in animals, particularly mastitis, which is treated in a dedicated chapter; when such foods are ingested raw or insufficiently cooked, staphylococcal proliferation and toxin release may result in disease.[79–85]

As mentioned, outbreaks of *B. cereus*, *S. aureus*, and *C. perfringens* probably suffer from greater underreporting owing to the usually mild illnesses they cause, which result in few patients seeking medical care; again, nondiagnosis may affect proper reporting because of the lack of routine clinical testing for these pathogens and related toxins from stool samples.[79–85]

Current public health interventions should aim at reducing contamination at the origin of the food production chain as well as cross-contamination in the food preparation areas; a further purpose should be educating the general public, mostly food handlers, as to the importance of preventing temperature abuse and improving food-handling procedures.

## REFERENCES

1. Hong HA, Duc Le H, Cutting SM. The use of bacterial spore formers as probiotics. *FEMS Microbiol Rev* 2005;**29**:813–35.
2. Mahler H, Pasi A, Kramer JM, Schulte P, Scoging AC, Bär W, et al. Fulminant liver failure in association with the emetic toxin of *Bacillus cereus*. *N Engl J Med* 1997;**336**:1142–8.
3. Lund T, De Buyser ML, Granum PE. A new cytotoxin from *Bacillus cereus* that may cause necrotic enteritis. *Mol Microbiol* 2000;**38**:254–61.
4. Dierick K, Van Coillie E, Swiecicka I, Meyfroidt G, Devlieger H, Meulemans A, et al. Fatal family outbreak of *Bacillus cereus*-associated food poisoning. *J Clin Microbiol* 2005;**43**:4277–9.
5. Ehling-Schulz M, Fricker M, Scherer S. *Bacillus cereus*, the causative agent of an emetic type of food-borne illness. *Mol Nutr Food Res* 2004;**48**:479–87.

6. Beecher DJ, MacMillan JD. Characterization of the components of hemolysin BL from *Bacillus cereus*. *Infect Immun* 1991;**59**:1778–84.

7. Lund T, Granum PE. Characterisation of a nonhaemolytic enterotoxin complex from *Bacillus cereus* isolated after a foodborne outbreak. *FEMS Microbiol Lett* 1996;**141**:151–6.

8. Kramer JM, Gilbert RJ. *Bacillus cereus* and other *Bacillus* species. In: Doyle MP, editor. *Foodborne bacterial pathogens*. New York: Marcel Dekker; 1989. p. 21–70.

9. Johnson KM. *Bacillus cereus* food-borne illness. An update. *J Food Prot* 1984;**47**:145–53.

10. Gilbert RJ, Kramer JM. *Bacillus cereus* food poisoning. In: Cliver DC, Cochrane BA, editors. *Progress in food safety (proceedings of symposium)*. Madison (WI): Food Research Institute, University of Wisconsin-Madison; 1986. p. 85–93.

11. Granum PE. Bacillus cereus. In: Doyle MP, Beuchat LR, editors. *Food microbiology: fundamentals and frontiers*. Washington (DC): ASM Press; 2007. p. 445–55.

12. Andersson A, Rönner U, Granum PE. What problems does the food industry have with the spore-forming pathogens *Bacillus cereus* and *Clostridium perfringens*? *Int J Food Microbiol* 1995;**28**:145–55.

13. Wiencek KM, Klapes NA, Foegeding PM. Hydrophobicity of *Bacillus* and *Clostridium* spores. *Appl Environ Microbiol* 1990;**56**:2600–5.

14. Tauveron G, Slomianny C, Henry C, Faille C. Variability among *Bacillus cereus* strains in spore surface properties and influence on their ability to contaminate food surface equipment. *Int J Food Microbiol* 2006;**110**:254–62.

15. Faille C, Tauveron G, Le Gentil-Lelievre C, Slomianny C. Occurrence of *Bacillus cereus* spores with a damaged exosporium: consequences on the spore adhesion on surfaces of food processing lines. *J Food Prot* 2007;**70**:2346–53.

16. Faille C, Jullien C, Fontaine F, Bellon-Fontaine MN, Slomianny C, Benezech T. Adhesion of *Bacillus* spores and *Escherichia coli* cells to inert surfaces: role of surface hydrophobicity. *Can J Microbiol* 2002;**48**:728–38.

17. Wijman JG, de Leeuw PP, Moezelaar R, Zwietering MH, Abee T. Air-liquid interface biofilms of *Bacillus cereus*: formation, sporulation, and dispersion. *Appl Environ Microbiol* 2007;**73**:1481–8.

18. Ryu JH, Beuchat LR. Biofilm formation and sporulation by *Bacillus cereus* on a stainless steel surface and subsequent resistance of vegetative cells and spores to chlorine, chlorine dioxide, and a peroxyacetic acid-based sanitizer. *J Food Prot* 2005;**68**:2614–22.

19. Wong HC, Chang MH, Fan JY. Incidence and characterization of *Bacillus cereus* isolates contaminating dairy products. *Appl Environ Microbiol* 1988;**54**:699–702.

20. TeGiffel MC, Beumer RR, Granum PE, Rombouts FM. Isolation and characterisation of *Bacillus cereus* from pasteurised milk in household refrigerators in The Netherlands. *Int J Food Microbiol* 1977;**34**:307–18.

21. Larsen HD, Jørgensen K. Growth of *Bacillus cereus* in pasteurized milk products. *Int J Food Microbiol* 1999;**46**:173–6.

22. Granum PE. *Bacillus cereus* and its toxins. *Soc Appl Bacteriol Symp Ser* 1994;**23**:61S–6S.

23. Hauge S. Food poisoning caused by *Bacillus cereus*. *Nord Hyg Tidsskr* 1950;**31**:189–206.

24. Hauge S. Food poisoning caused by aerobic spore forming bacilli. *J Appl Bacteriol* 1955;**18**:591–5.

25. Clavel T, Carlin F, Lairon D, Nguyen-The C, Schmitt P. Survival of *Bacillus cereus* spores and vegetative cells in acid media simulating human stomach. *J Appl Microbiol* 2004;**97**:214–9.

26. Mortimer PR, McCann G. Food-poisoning episodes associated with *Bacillus cereus* in fried rice. *Lancet* 1974;**1**:1043–5.

27. Seo KS, Bohach GA. Staphylococcus aureus. In: Doyle MP, Beuchat LR, editors. *Food microbiology: fundamentals and frontiers*. Washington (DC): ASM Press; 2007. p. 493–518.

28. Jääskeläinen EL, Teplova V, Andersson MA, Andersson LC, Tammela P, Andersson MC, et al. In vitro assay for human toxicity of cereulide, the emetic mitochondrial toxin produced by food poisoning *Bacillus cereus*. *Toxicol In Vitro* 2003;**17**:737–44.

29. Fricker M, Messelhäusser U, Busch U, Scherer S, Ehling-Schulz M. Diagnostic real-time PCR assays for the detection of emetic *Bacillus cereus* strains in foods and recent food-borne outbreaks. *Appl Environ Microbiol* 2007;**73**:1892–8.

30. Granum PE, Brynestad S, O'Sullivan K, Nissen H. Enterotoxin from *Bacillus cereus*: production and biochemical characterization. *Neth Milk Dairy J* 1993;**47**:63–70.

31. Andersson A, Granum PE, Rönner U. The adhesion of *Bacillus cereus* spores to epithelial cells might be an additional virulence mechanism. *Int J Food Microbiol* 1998;**39**:93–9.

32. Granum PE. *Clostridium perfringens* toxins involved in food poisoning. *Int J Food Microbiol* 1990;**10**:101–11.

33. Kotiranta A, Lounatmaa K, Haapasalo M. Epidemiology and pathogenesis of *Bacillus cereus* infections. *Microbes Infect* 2000;**2**:189–98.

34. Schoeni JL, Wong AC. *Bacillus cereus* food poisoning and its toxins. *J Food Prot* 2005;**68**:636–48.

35. Shinagawa K, Konuma H, Sekita H, Sugii S. Emesis of rhesus monkeys induced by intragastric administration with the HEp-2 vacuolation factor (cereulide) produced by *Bacillus cereus*. *FEMS Microbiol Lett* 1995;**130**:87–90.

36. Clavel T, Carlin F, Dargaignaratz C, Lairon D, Nguyen-The C, Schmitt P. Effects of porcine bile on survival of *Bacillus cereus* vegetative cells and Haemolysin BL enterotoxin production in reconstituted human small intestine media. *J Appl Microbiol* 2007;**103**:1568–75.

37. Stenfors Arnesen LP, Fagerlund A, Granum PE. From soil to gut: *Bacillus cereus* and its food poisoning toxins. *FEMS Microbiol Rev* 2008;**32**:579–606.

38. van Wely KH, Swaving J, Freudl R, Driessen AJ. Translocation of proteins across the cell envelope of Gram positive bacteria. *FEMS Microbiol Rev* 2001;**25**:437–54.

39. Ghelardi E, Celandroni F, Salvetti S, Beecher DJ, Gominet M, Lereclus D, et al. Requirement of *flhA* for swarming differentiation, flagellin export, and secretion of virulence-associated proteins in *Bacillus thuringiensis*. *J Bacteriol* 2002;**184**:6424–33.

40. Ghelardi E, Celandroni F, Salvetti S, Ceragioli M, Beecher DJ, Senesi S, et al. Swarming behavior and hemolysin BL secretion in *Bacillus cereus*. *Appl Environ Microbiol* 2007;**73**:4089–93.

41. Salvetti S, Ghelardi E, Celandroni F, Ceragioli M, Giannessi F, Senesi S. FlhF, a signal recognition particle-like GTPase, is involved in the regulation of flagellar arrangement, motility behaviour and protein secretion in *Bacillus cereus*. *Microbiology* 2007;**153**(Pt 8): 2541–52.

42. Bouillaut L, Ramarao N, Buisson C, Gilois N, Gohar M, Lereclus D, et al. FlhA influences *Bacillus thuringiensis* PlcR-regulated gene transcription, protein production, and virulence. *Appl Environ Microbiol* 2005;**71**:8903–10.

43. Stenfors LP, Mayr R, Scherer S, Granum PE. Pathogenic potential of fifty *Bacillus weihenstephanensis* strains. *FEMS Microbiol Lett* 2002;**215**:47–51.

44. Lechner S, Mayr R, Francis KP, Prüss BM, Kaplan T, Wiessner-Gunkel E, et al. *Bacillus weihenstephanensis* sp. nov. is a new psychrotolerant species of the *Bacillus cereus* group. *Int J Syst Bacteriol* 1998;**48**:1373–82.

45. Agata N, Ohta M, Yokoyama K. Production of *Bacillus cereus* emetic toxin (cereulide) in various foods. *Int J Food Microbiol* 2002;**73**:23–7.

46. Bauer T, Stark T, Hofmann T, Ehling-Schulz M. Development of a stable isotope dilution analysis for the quantification of the *Bacillus cereus* toxin cereulide in foods. *J Agric Food Chem* 2010;**58**:1420–8.

47. Bottone EJ. *Bacillus cereus*, a volatile human pathogen. *Clin Microbiol Rev* 2010;**23**:382–98.

48. Cardazzo B, Negrisolo E, Carraro L, Alberghini L, Patarnello T, Giaccone V. Multiple-locus sequence typing and analysis of toxin genes in *Bacillus cereus* food-borne isolates. *Appl Environ Microbiol* 2008;**74**:850–60.

49. Ceuppens S, Boon N, Rajkovic A, Heyndrickx M, Van de Wiele T, Uyttendaele M. Quantification methods for *Bacillus cereus* vegetative cells and spores in the gastrointestinal environment. *J Microbiol Methods* 2010;**83**:202–10.

50. Chaves JQ, Pires ES, Vivoni AM. Genetic diversity, antimicrobial resistance and toxigenic profiles of *Bacillus cereus* isolated from food in Brazil over three decades. *Int J Food Microbiol* 2011;**147**:12–6.

51. Choma C, Granum PE. The enterotoxin T (BcET) from *Bacillus cereus* can probably not contribute to food poisoning. *FEMS Microbiol Lett* 2002;**217**:115–9.

52. De Jonghe V, Coorevits A, De Block J, Van Coillie E, Grijspeerdt K, Herman L, et al. Toxinogenic and spoilage potential of aerobic spore-formers isolated from raw milk. *Int J Food Microbiol* 2010;**136**:318–25.

53. Dzieciol M, Fricker M, Wagner M, Hein I, Ehling-Schulz M. A diagnostic real-time PCR assay for quantification and differentiation of emetic and non-emetic *Bacillus cereus* in milk. *Food Control* 2013;**32**:176–85.

54. Ehling-Schulz M, Fricker M, Scherer S. Identification of emetic toxin producing *Bacillus cereus* strains by a novel molecular assay. *FEMS Microbiol Lett* 2004;**232**:189–95.

55. Ehling-Schulz M, Guinebretiere MH, Monthan A, Berge O, Fricker M, Svensson B. Toxin gene profiling of enterotoxic and emetic *Bacillus cereus*. *FEMS Microbiol Lett* 2006;**260**:232–40.

56. Ehling-Schulz M, Svensson B, Guinebretiere MH, Lindback T, Andersson M, Schulz A, et al. Emetic toxin formation of *Bacillus cereus* is restricted to a single evolutionary lineage of closely related strains. *Microbiology* 2005;**151**(Pt 1):183–97.

57. Elhanany E, Barak R, Fisher M, Kobiler D, Altboum Z. Detection of specific *Bacillus anthracis* spore biomarkers by matrix assisted laser desorption/ionization time-of-flight mass spectrometry. *Rapid Commun Mass Spectrom* 2001;**15**:2110–6.

58. Fagerlund A, Lindback T, Storset AK, Granum PE, Hardy SP. *Bacillus cereus* Nhe is a pore-forming toxin with structural and functional properties similar to the ClyA (HlyE, SheA) family of haemolysins, able to induce osmotic lysis in epithelia. *Microbiology* 2008;**154**(Pt 3):693–704.

59. Fagerlund A, Ween O, Lund T, Hardy SP, Granum PE. Genetic and functional analysis of the cytK family of genes in *Bacillus cereus*. *Microbiology* 2004;**150**(Pt 8):2689–97.

60. Guinebretiere MH, Broussolle V, Nguyen-The C. Enterotoxigenic profiles of food-poisoning and food-borne *Bacillus cereus* strains. *J Clin Microbiol* 2002;**40**:3053–6.

61. Guinebretière MH, Velge P, Couvert O, Carlin F, Debuyser ML, Nguyen-The C. Ability of *Bacillus cereus* group strains to cause food poisoning varies according to phylogenetic affiliation (groups I to VII) rather than species affiliation. *J Clin Microbiol* 2010;**48**:3388–91.

62. Jackson SG, Goodbrand RB, Ahmed R, Kasatiya S. *Bacillus cereus* and *Bacillus thuringiensis* isolated in a gastroenteritis outbreak investigation. *Lett Appl Microbiol* 1995;**21**:103–5.

63. Kumar TD, Murali HS, Batra HV. Construction of a non toxic chimeric protein (L1-L2-B) of Haemolysin BL from *Bacillus cereus* and its application in HBL toxin detection. *J Microbiol Methods* 2008;**75**:472–7.

64. Martínez-Blanch JF, Sánchez G, Garay E, Aznar R. Development of a real-time PCR assay for detection and quantification of enterotoxigenic members of *Bacillus cereus* group in food samples. *Int J Food Microbiol* 2009;**35**:15–21.

65. Mikkola R, Saris NE, Grigoriev PA, Andersson MA, Salkinoja-Salonen MS. Ionophoretic properties and mitochondrial effects of cereulide: the emetic toxin of *Bacillus cereus*. *Eur J Biochem* 1999;**263**:112–7.

66. Ngamwongsatit P, Buasri W, Pianariyanon P, Pulsrikarn C, Ohba M, Assavanig A, et al. Broad distribution of enterotoxin genes (hblCDA, nheABC, cytK, and entFM) among *Bacillus thuringiensis* and *Bacillus cereus* as shown by novel primers. *Int J Food Microbiol* 2008;**121**:352–6.

67. Paananen A, Mikkola R, Sareneva T, Matikainen S, Hess M, Andersson M, et al. Inhibition of human natural killer cell activity by cereulide, an emetic toxin from *Bacillus cereus*. *Clin Exp Immunol* 2002;**129**:420–8.

68. Peng H, Ford V, Frampton EW, Restaino L, Shelef LA, Spitz H. Isolation and enumeration of *Bacillus cereus* from foods on a novel chromogenic plating media. *Food Microbiol* 2001;**18**:231–8.
69. Swaminathan B, Gerner-Smidt P, Ng LK, Lukinmaa S, Kam KM, Rolando S, et al. Building PulseNet International: an interconnected system of laboratory networks to facilitate timely public health recognition and response to foodborne disease outbreaks and emerging foodborne diseases. *Foodborne Pathog Dis* 2006;**3**:36–50.
70. Tsilia V, Devreese B, de Baenst I, Mesuere B, Rajkovic A, Uyttendaele M, et al. Application of MALDI-TOF mass spectrometry for the detection of enterotoxins produced by pathogenic strains of the *Bacillus cereus* group. *Anal Bioanal Chem* 2012;**404**:1691–702.
71. Wehrle E, Didier A, Moravek M, Dietrich R, Märtlbauer E. Detection of *Bacillus cereus* with enteropathogenic potential by multiplex real-time PCR based on SYBR Green I. *Mol Cell Probes* 2010;**24**:124–30.
72. Yamada S, Ohashi E, Agata N, Venkateswaran K. Cloning and nucleotide sequence analysis of *gyrB* of *Bacillus cereus*, *B. thuringiensis*, *B. mycoides*, and *B. anthracis* and their application to the detection of *B. cereus* in rice. *Appl Environ Microbiol* 1999;**65**:1483–90.
73. Baron F, Cochet MF, Grosset N, Madec MN, Briandet R, Dessaigne S, et al. Isolation and characterization of a psychrotolerant toxin producer, *Bacillus weihenstephanensis*, in liquid egg products. *J Food Prot* 2007;**70**:2782–91.
74. Agata N, Mori M, Ohta M, Suwan S, Ohtani I, Isobe M. A novel dodecadepsipeptide, cereulide, isolated from *Bacillus cereus* causes vacuole formation in HEp-2 cells. *FEMS Microbiol Lett* 1994;**121**:31–4.
75. Agata N, Ohta M, Mori M, Isobe M. A novel dodecadepsipeptide, cereulide, is an emetic toxin of *Bacillus cereus*. *FEMS Microbiol Lett* 1995;**129**:17–20.
76. Agata N, Ohta M, Mori M, Shibayama K. Growth conditions of and emetic toxin production by *Bacillus cereus* in a defined medium with amino acids. *Microbiol Immunol* 1999;**43**:15–8.
77. Andersson MA, Mikkola R, Helin J, Andersson MC, Salkinoja-Salonen M. A novel sensitive bioassay for detection of *Bacillus cereus* emetic toxin and related depsipeptide ionophores. *Appl Environ Microbiol* 1998;**64**:1338–43.
78. Jiménez G, Urdiain M, Cifuentes A, López-López A, Blanch AR, Tamames J, et al. Description of *Bacillus toyonensis* sp. nov., a novel species of the *Bacillus cereus* group, and pairwise genome comparisons of the species of the group by means of ANI calculations. *Syst Appl Microbiol* 2013;**36**:383–91.
79. Bennett SD, Walsh KA, Gould LH. Foodborne disease outbreaks caused by *Bacillus cereus*, *Clostridium perfringens*, and *Staphylococcus aureus* - United States, 1998–2008. *Clin Infect Dis* 2013;**57**:425–33.
80. Stewart GC. *Staphylococcus aureus*. In: Fratamico P, Bhunia A, Smith J, editors. *Foodborne pathogens: microbiology and molecular biology*. Norfolk (UK): Caister Academic Press; 2005. p. 273–84.
81. Le Loir Y, Baron F, Gautier M. *Staphylococcus aureus* and food poisoning. *Genet Mol Res* 2003;**2**:63–76.
82. Haeghebaert S, Le Querrec F, Gallay A, Bouvet P, Gomez M, Vaillant V. Food poisonings in France, 1999–2000. *BEH* 2002;**23**:105–9.
83. Scallan E, Jones TF, Cronquist A, et al. Factors associated with seeking medical care and submitting a stool sample in estimating the burden of foodborne illness. *Foodborne Pathog Dis* 2006;**3**:432–8.
84. Guerrant RL, Van Gilder T, Steiner TS, et al. Practice guidelines for the management of infectious diarrhea. *Clin Infect Dis* 2001;**32**:331.
85. Thielman NM, Guerrant RL. Acute infectious diarrhea. *N Engl J Med* 2004;**350**:38–47.

# CHAPTER 6

# *Bacillus cereus* Pneumonia

Vincenzo Savini
Clinical Microbiology and Virology, Laboratory of Bacteriology and Mycology, Civic Hospital of Pescara, Pescara, Italy

## SUMMARY

*Bacillus cereus* is a ubiquitously distributed environmental organism whose pathogenicity for airways is probably poorly taken into consideration in the medical setting but emerges from the published literature. Notably, in fact, the organism does not always behave as a mere contaminant when cultivated from human airway samples, but potentially causes lung infection that may be severe and mimic anthrax. This chapter therefore aims to emphasize the role of this bacterium as a respiratory tract pathogen and highlight virulence and antibiotic-resistance determinants that contribute to making *B. cereus* pneumonia a potentially life-threatening condition.

## BACKGROUND

*Bacillus anthracis* is the major member of the genus *Bacillus*, as it is notoriously a frank pathogen for skin and gut and, above all, for airways, where it causes the so-called, unfortunately famous and often fatal, "anthrax."[1,36,39,42,43,48,56,57,61,62]

*Bacillus cereus* is instead mostly an environmental spore-forming organism, almost ubiquitously distributed in nature, whose reservoir is represented by soil, decaying organic matter, vegetation, fresh and marine waters, and the invertebrate gut (from which soil and food, and subsequently the human enteric tract, may become contaminated), along with dirt, air, and stools.[26,41,54,60,72,80,83,87,90] Particularly, among plants, the bacterium has been frequently found as a rice contaminant, so food-related intoxication caused by this organism's toxins is known to be frequently associated with the ingestion of this product, as described in a dedicated chapter of this book.[14]

Given the ubiquity of *B. cereus*, the pathogenicity it may display is not always acknowledged when the organism is isolated from human clinical materials; most commonly, in fact, it is labeled as a harmless contaminant and dismissed, even aprioristically.[42]

*The Diverse Faces of Bacillus cereus*
ISBN 978-0-12-801474-5
http://dx.doi.org/10.1016/B978-0-12-801474-5.00006-2

*Bacillus cereus* should not be considered as a single bacterium, as it is part (*B. cereus sensu stricto*) of a group of environmental organisms, named as the *B. cereus* group (otherwise *B. cereus sensu lato*), which includes the eight species *B. anthracis* (the above-mentioned agent of anthrax), *B. cereus sensu stricto* (mostly responsible for an emetic and a diarrheal food-related syndrome), *Bacillus mycoides*, *Bacillus pseudomycoides*, *Bacillus thuringiensis* (which has been known to be a natural insecticide source), *Bacillus weihenstephanensis*, *Bacillus cytotoxicus*, and *Bacillus toyonensis*.[13,19,20,22,23,28,29,47,51,53,59,65,67,79,81,84,86,93–96,105]

As said, *B. cereus* is a ubiquitous environmental organism. Notably, being a sporulating bacterium, its survival in the environment is strictly related to formation of spores, which are resistant to extreme conditions such as drying, heat, freezing, and radiation and are the true infective form of this pathogen.[12]

In the food industry, spores are of particular concern as they can tolerate pasteurization as well as γ-radiation; furthermore, their hydrophobicity makes them adhere to surfaces and then spread to almost all kinds of food.[12,44,52,92,98,102] Given the ubiquity of *B. cereus* in food products, then, it is ingested and subsequently and transiently becomes a part of the enteric microbiota in humans; it is not yet understood, however, if isolation of this organism from stools occurs by virtue of germinating spores or growth of vegetative cells.[12]

## PATHOLOGY

*Bacillus cereus* is not just a harmless contaminant in clinical laboratories but has been widely described to be an agent of human diseases that are mostly food-borne intoxications[2,3,9,18,30,32,34,35,50,54,70,74,74,76]; in particular, these are, usually, self-limited illnesses that occur in the form of either an emetic syndrome (owing to an emetic toxin and often following ingestion of fried rice) or a diarrheal one (which relies on the production of enterotoxins).[14,33–35,41,42,55,69,77,88] Typically, when large quantities of rice are cooked and then get cold slowly, conditions that predispose to spore germination may be established; consequently, toxin is released into the food, before ingestion, or into the gut, after those foods are eaten.[14]

Together with such above-mentioned food-related clinical syndromes, however, *B. cereus* has emerged in the past decades as an agent of several pathologic processes, including ocular infections (that is, endophthalmitis, panophthalmitis, or keratitis, usually based on a traumatic event), periodontal infections, skin and postsurgical wound infections (even with bone involvement), osteomyelitis, necrotizing fasciitis, meningitis, endocarditis, salpingitis, and bacteremia.[14,31,42,54,60,83,99,100]

Patients that suffer from an immune system impairment, such as those with hematologic malignancies, preterm newborns, critically ill and debilitated subjects, and people recovering from surgical procedures, are particularly prone to *B. cereus* infections, and, especially, sepsis, sepsis–related multiorgan failure, and unfavorable clinical outcomes are unfortunately and frequently observed in the event of an underlying neutropenia.[42,45,54,58,60] Nevertheless, immunocompetent people may be affected by this organism, as well.[17]

Notably, *B. cereus* has been unearthed as an uncommon agent of lung infection, which may be serious, as it sometimes looks like the famous anthrax and mostly involves immunocompromised hosts, who have more than once been led to the exitus.[11,27,40,66,89]

Otherwise healthy people can, however, develop *B. cereus* pneumonia as well, welders being a category of subjects showing an increased risk of acquiring such an infection by *B. cereus* by virtue of poorly understood reasons.[54,72] Of interest, in this context, pulmonary infections mimicking anthrax have been described to be due to *B. cereus* strains that harbored *B. anthracis* toxin genes.[12]

Based on current knowledge, *B. cereus* pneumonia patients are usually referred and/or present to medical attention with dyspnea, productive cough, and hemoptysis, with or without pleural effusions or empyema.[6,17,72,104] Pneumonia-associated presentations outside the respiratory tract have been reported, too, including bullous skin lesions, nausea and vomiting, hematemesis, and diarrhea, together with fever, chills, leukocytosis, and bacteremia. Generally, unfortunately, the disease is serious and the clinical outcome often unfavorable.[6,17,72,104]

Owing to the frequent lethality of *B. cereus* lung infections, most of the available histopathological data rely on postmortem investigations. Autopsies have in fact revealed necrotizing pneumonias, bronchopneumonias, presence of large amounts of rods inside alveoli and pleura, abundant serosanguinous fluid occupying the pleural cavities, fibrinopurulent material adherent to the pleural surface, and hemorrhage and edema of both the pleura and the interlobular septa, along with intra-alveolar edema and flogosis.[6,17,104] As extralung findings, postmortem studies have instead documented congestion of the spleen, together with presence of bacilli in the serous fluid overflowing from the cutaneous bullous lesions.[104]

A particular and unique case, as emerges from the published literature, is that of a fatal tracheobronchitis with pneumonia described in 2001 in an aplastic anemia patient, who initially suffered from chest pain and yellowish sputum, but the infection then quickly evolved to a fatal, anthrax-like pneumonia.[90] On examination, a severe pseudomembranous flogosis of the trachea and the bronchial tree was observed, as well as a diffuse alveolar damage and hemorrhage. Again, fiber-optic bronchoscopy showed the presence of a strongly inflamed mucosa in the trachea and the bronchi and a bilateral obstruction of the entire visible bronchial tree caused by diphtheria-like membranes that could not be removed owing to mucosal bleeding.[90]

In general, *B. cereus* shows propensity to cause necrotizing processes,[90] and although pneumonia has been mostly found to be sporadic, nosocomial outbreaks have occurred, as well, owing to contaminated hospital linen or inadequately sterilized respiratory circuits.[60] Given the ubiquitous presence of this organism in the environment, however, it is again hard to understand with certainty the pathogenic role of *B. cereus* in the context of a lung illness, even with a human respiratory sample.[60]

## PATHOGENICITY

*Bacillus cereus* pathogenicity is intimately associated with the synthesis of tissue-destructive exoenzymes, including phospholipases, an emesis-inducing toxin (causing the vomiting syndrome), and the pore-forming enterotoxins cytotoxin K, hemolysin BL, and nonhemolytic enterotoxin.[4,5,7,8,10,12,16, 21,24,37,38,46,49,63,64,68,71,75,78,85,90,97]

*Bacillus cereus* is genotypically strictly related to *B. anthracis*[104]; in fact, *B. cereus* isolates that share genome similarity with *B. anthracis* are mostly of clinical rather than environmental origin and may sometimes harbor *B. anthracis* virulence genes.[54] These encode extracellular molecules (the capsule, the lethal factor (LF), the edema factor (EF), and the protective antigen (PA)) that are involved in the mentioned *B. cereus* clinical syndrome resembling inhalation anthrax.[104] In particular, *B. cereus* strains that produce the anthrax-related toxin genes *pagA*, *lef*, and *cya* (encoding PA, LF, and EF, respectively) have been responsible for fatal cases of pneumonia.[6,41,104]

Conversely, in other cases of disease, it has been found that plasmids carrying *B. anthracis* toxin genes seem not to be required for development of severe pneumonia, as certain isolates from severe clinical episodes did not harbor plasmid pXO1 (harboring the *pagA*, *lef*, and *cya* genes).[12]

The *B. anthracis* capsule is encoded by genes *capA*, *capB*, and *capC* in the pXO2 virulence plasmid and is essential for the development of anthrax; likewise, certain *B. cereus* strains produce a capsule that plays a role in the pathogenesis of inhalation anthrax-like *B. cereus* disease.[91,104] As an example, *B. cereus* strains G9241, 03BB87, and 03BB102, which have been responsible for anthrax-like pneumonias, share relevant genome homology with *B. anthracis* and harbor almost the entire pXO1 plasmid of virulence, which is in its turn related to anthrax. Nonetheless, while the *B. anthracis* capsule is composed of poly-D-γ-glutamic acid, strains G9241, 03BB87, and 03BB102 capsules show a polysaccharide composition.[41,54,104]

Moreover, strains G9241 and 03BB87 carry the plasmid pBC218, which is probably involved in capsule formation[41]; strain 03BB102 instead harbors genes *capA*, *capB*, and *capC* (on which *B. anthracis* capsule synthesis relies), but this strain's capsule is made of polysaccharides, similar to strains G9241 and 03BB87.[54]

In anthrax, the disease pathogenesis involves phagocytosis of spores by lung dendritic cells and then spore transport within these cells as far as the tracheobronchial lymph nodes, where spores germinate. The following steps are replication of vegetative bacteria, ET and LT secretion, and subsequent development of extensive tissue damage, hemorrhage, extension of the pathologic process through the mediastinum, dissemination via the circulation, and hemorrhagic meningitis, which represents the major cause of exitus.[101] Likewise, it is possible that toxin-producing *B. cereus* strains adopt the same (or a similar) mechanism leading to tissue necrosis and organ impairment.[60]

## ANTIBIOTICS

Therapeutic strategies against *B. cereus* infection should be based on a documented antibiotic susceptibility profile, but unfortunately species-specific criteria for evaluation of in vitro antibiotic activity are still lacking. Most *B. cereus* isolates are resistant to penicillins and cephalosporins by virtue of production of a β-lactamase, and resistance of clinical isolates to these compounds should be considered nowadays to be constant.[15,25]

Even a chromosomal metallo-β-lactamase (MBL) is widespread among *B. cereus* strains (the enzyme is the so-called "BcII")[15] and, notably, *B. cereus* MBL was the first to be discovered within this category of enzymes, in 1966.[82] Only later, in fact, were MBLs found in *Stenotrophomonas maltophilia*, *Aeromonas* spp., *Pseudomonas aeruginosa*, *Acinetobacter* spp., and *Bacteroides fragilis*. MBLs are known to inhibit all β-lactams except for aztreonam (Gram-positive organisms like *B. cereus* are, however, constitutively resistant to the latter).[83] Finally *B. cereus* commonly produces Bush group 2a penicillinases I and III, which hydrolyze penicillins and are inhibited by clavulanic acid, which may be therefore of therapeutic support against those strains that do not produce MBLs.[83]

Within the non-β-lactam antibiotic drugs, *B. cereus* may be resistant to cotrimoxazole, clindamycin, erythromycin, and tetracycline,[12,15,42,60,72,83,104] while most strains seem to respond to daptomycin and linezolid.[12] Combined vancomycin and daptomycin resistance has nevertheless been described, unfortunately, and this information seems to suggest the existence of differences in bacterial membrane composition of those mentioned organisms if compared with vancomycin- and daptomycin-susceptible strains.[15]

In light of this, it is challenging to foresee *B. cereus* behavior under antimicrobial exposure, but it is clear that broad-spectrum cephalosporins should be avoided as an empirical treatment of pneumonia when *B. cereus* is reasonably the etiologic agent of it.[12]

## FINAL REMARKS

Outside the notoriety of *B. cereus* in the field of food poisoning, attention should be given to this organism as an agent of pneumonia.[12]

There is no conclusive explanation, thus far, for the strong relationship existing between the professional category of metal workers and *B. cereus* lung infection.[12,54] Published works seem to suggest that welders may suffer with higher frequencies from pneumonias showing enhanced seriousness and duration, but indeed, it is impossible to link such clinical episodes with any environmental sources of *B. cereus*.[6,54]

Colonization of the oral cavity is instead crucial in hosts with an underlying compromised immune system in the background and it may occur either through the inhalation of spores or by vegetative bacteria; then, *B. cereus* may spread to adjacent tissues or disseminate through the blood circulation.[12] In the unique pseudomembranous tracheobronchitis case described, particularly, it is likely that drug-related damage of the oral mucosa had enhanced spore and vegetative cell adherence to the epithelium and then subsequent colonization.[12,90] Interestingly, that pseudomembranous tracheobronchitis case looked like diphtheria, as pseudomembranes in the respiratory tract are typically related to *Corynebacterium diphtheriae* disease; nevertheless, pseudomembranous tracheobronchitis caused by *Aspergillus* species and corynebacteria other than *C. diphtheriae* have been described in recent years and were found to involve immunocompromised patients.[90]

Finally, it has been suggested that asthma and smoking could predispose to *B. cereus* lung infection, but their role, indeed, is still poorly understood.[6]

Pulmonary anthrax is mostly observed in livestock and less commonly in humans.[41,72] It is life-threatening and usually accompanied by overwhelming sepsis. The syndrome starts abruptly with fever, dyspnea, and chest pain; it progresses quickly and frequently brings patients to death before the antibiotic treatment may produce any results.[72,73,103] Similarly, in anthrax-like *B. cereus* pneumonia, it is likely that the great amount of *B. cereus* cells along with the wide spectrum of enzymes and toxins produced may justify the rapid course and the final unfavorable outcome.[72]

To conclude, it emerges from the published literature that knowledge of the airway pathogenicity of *B. cereus* is still very unclear, and future microbiological, clinical, and histopathologic data will shed further light on this topic. In clinical laboratories, quick diagnostic assays are required that discriminate virulent strains from the vast majority of contaminant isolates[6]; hopefully, awareness of *B. cereus* pneumonia both as a professional disease and as a matter of public health interest will get increasing attention in the near future.

## REFERENCES

1. Abshire TG, Brown JE, Ezzell JW. Production and validation of the use of gamma phage for identification of *Bacillus anthracis*. *J Clin Microbiol* 2005;**43**:4780–8.
2. Agata N, Mori M, Ohta M, Suwan S, Ohtani I, Isobe M. A novel dodecadepsipeptide, cereulide, isolated from *Bacillus cereus* causes vacuole formation in HEp-2 cells. *FEMS Microbiol Lett* 1994;**121**:31–4.
3. Agata N, Ohta M, Yokoyama K. Production of *Bacillus cereus* emetic toxin (cereulide) in various foods. *Int J Food Microbiol* 2002;**73**:23–7.
4. Andreeva ZI, Nesterenko VF, Fomkina MG, Ternovsky VI, Suzina NE, Bakulina AY, et al. The properties of *Bacillus cereus* hemolysin II pores depend on environmental conditions. *Biochim Biophys Acta* 2007;**1768**:253–63.
5. Andreeva ZI, Nesterenko VF, Yurkov IS, Budarina ZI, Sineva EV, Solonin AS. Purification and cytotoxic properties of *Bacillus cereus* hemolysin II. *Protein Expr Purif* 2006;**47**:186–93.
6. Avashia SB, Riggins WS, Lindley C, Hoffmaster A, Drumgoole R, Nekomoto T, et al. Fatal pneumonia among metalworkers due to inhalation exposure to *Bacillus cereus* containing *Bacillus anthracis* toxin genes. *Clin Infect Dis* 2007;**44**:414–6.
7. Baida G, Budarina ZI, Kuzmin NP, Solonin AS. Complete nucleotide sequence and molecular characterization of hemolysin II gene from *Bacillus cereus*. *FEMS Microbiol Lett* 1999;**180**:7–14.
8. Baida GE, Kuzmin NP. Mechanism of action of hemolysin III from *Bacillus cereus*. *Biochim Biophys Acta* 1996;**1284**:122–4.
9. Bauer T, Stark T, Hofmann T, Ehling-Schulz M. Development of a stable isotope dilution analysis for the quantification of the *Bacillus cereus* toxin cereulide in foods. *J Agric Food Chem* 2010;**58**:1420–8.

10. Beecher DJ, Wong AC. Cooperative, synergistic and antagonistic haemolytic interactions between haemolysin BL, phosphatidylcholine phospholipase C and sphingomyelinase from *Bacillus cereus*. *Microbiology* 2000;**146**(Pt 12):3033–9.

11. Bekemeyer WB, Zimmerman GA. Life-threatening complications associated with *Bacillus cereus* pneumonia. *Am Rev Respir Dis* 1985;**131**:466–9.

12. Bottone EJ. *Bacillus cereus*, a volatile human pathogen. *Clin Microbiol Rev* 2010;**23**:382–98.

13. Bravo A, Likitvivatanavong S, Gill SS, Soberón M. *Bacillus thuringiensis*: a story of a successful bioinsecticide. *Insect Biochem Mol Biol* 2011;**41**:423–31.

14. Brooks GF, Butel JS, Morse SA. *Medical microbiology*. 22nd ed. New York: McGraw-Hill; 2001.

15. Brown CS, Chand MA, Hoffman P, Woodford N, Livermore DM, Brailsford S, et al. United Kingdom incident response team. Possible contamination of organ preservation fluid with *Bacillus cereus*: the United Kingdom response. *Euro Surveill* 2012;**17**. pii:20165.

16. Budarina ZI, Nikitin DV, Zenkin N, Zakharova M, Semenova E, Shlyapnikov MG, et al. A new *Bacillus cereus* DNA-binding protein, HlyIIR, negatively regulates expression of *B. cereus* haemolysin II. *Microbiology* 2004;**150**(Pt 11):3691–701.

17. Carbone JE, Stauffer JL. *Bacillus cereus* pleuropulmonary infection in a normal host. *West J Med* 1985;**143**:676–7.

18. Cardazzo B, Negrisolo E, Carraro L, Alberghini L, Patarnello T, Giaccone V. Multiple-locus sequence typing and analysis of toxin genes in *Bacillus cereus* food-borne isolates. *Appl Environ Microbiol* 2008;**74**:850–60.

19. Carlson CR, Caugant DA, Kolstø AB. Genotypic diversity among *Bacillus cereus* and *Bacillus thuringiensis* strains. *Appl Environ Microbiol* 1994;**60**:1719–25.

20. Ceuppens S, Boon N, Rajkovic A, Heyndrickx M, Vande Wiele T, Uyttendaele M. Quantification methods for *Bacillus cereus* vegetative cells and spores in the gastrointestinal environment. *J Microbiol Methods* 2010;**83**:202–10.

21. Ceuppens S, Timmery S, Mahillon J, Uyttendaele M, Boon N. Small *Bacillus cereus* ATCC 14579 subpopulations are responsible for cytotoxin K production. *J Appl Microbiol* 2013;**114**:899–906.

22. Chaves JQ, Pires ES, Vivoni AM. Genetic diversity, antimicrobial resistance and toxigenic profiles of *Bacillus cereus* isolated from food in Brazil over three decades. *Int J Food Microbiol* 2011;**147**:12–6.

23. Chen ML, Tsen HY. Discrimination of *Bacillus cereus* and *Bacillus thuringiensis* with 16S rRNA and *gyrB* gene based PCR primers and sequencing of their annealing sites. *J Appl Microbiol* 2002;**92**:912–9.

24. Choma C, Granum PE. The enterotoxin T (BcET) from *Bacillus cereus* can probably not contribute to food poisoning. *FEMS Microbiol Lett* 2002;**217**:115–9.

25. Chon JW, Kim JH, Lee SJ, Hyeon JY, Seo KH. Toxin profile, antibiotic resistance and phenotypic and molecular characterization of *Bacillus cereus* in Sunsik. *Food Microbiol* 2012;**32**:217–22.

26. Colombo S, Carretto E. *Bacillus cereus* e specie correlate. In: Rondanelli EG, Fabbi M, Marone P, editors. *Trattato sulle infezioni e tossinfezioni alimentari*. Pavia (Italy): Selecta Medica; 2005. p. 609–39.

27. Coonrod JD, Leadley PJ, Eickhoff T.C.. *Bacillus cereus* pneumonia and bacteremia. A case report. *Am Rev Respir Dis* 1971;**103**:711–4.

28. Daffonchio D, Borin S, Frova G, Gallo R, Mori E, Fani R, et al. A randomly amplified poly morphic DNA marker specific for the *Bacillus cereus* group is diagnostic for *Bacillus anthracis*. *Appl Environ Microbiol* 1999;**65**:1298–303.

29. Daffonchio D, Cherif A, Borin S. Homoduplex and heteroduplex polymorphisms of the amplified ribosomal 16S-23S internal transcribed spacers describe genetic relationships in the "*Bacillus cereus* group". *Appl Env Microbiol* 2000;**66**:5460–8.

30. De Jonghe V, Coorevits A, De Block J, Van Coillie E, Grijspeerdt K, Herman L, et al. Toxinogenic and spoilage potential of aerobic spore-formers isolated from raw milk. *Int J Food Microbiol* 2010;**136**:318–25.

31. Dohmae S, Okubo T, Higuchi W, Takano T, Isobe H, Baranovich T, et al. *Bacillus cereus* nosocomial infection from reused towels in Japan. *J Hosp Infect* 2008;**69**:361–7.

32. Dzieciol M, Fricker M, Wagner M, Hein I, Ehling-Schulz M. A diagnostic real-time PCR assay for quantification and differentiation of emetic and non-emetic *Bacillus cereus* in milk. *Food Control* 2013;**32**:176–85.

33. Ehling-Schulz M, Fricker M, Scherer S. Identification of emetic toxin producing *Bacillus cereus* strains by a novel molecular assay. *FEMS Microbiol Lett* 2004;**232**:189–95.

34. Ehling-Schulz M, Guinebretiere MH, Monthan A, Berge O, Fricker M, Svensson B. Toxin gene profiling of enterotoxic and emetic *Bacillus cereus*. *FEMS Microbiol Lett* 2006;**260**:232–40.

35. Ehling-Schulz M, Svensson B, Guinebretiere MH, Lindback T, Andersson M, Schulz A, et al. Emetic toxin formation of *Bacillus cereus* is restricted to a single evolutionary lineage of closely related strains. *Microbiology* 2005;**151**(Pt 1):183–97.

36. Elhanany E, Barak R, Fisher M, Kobiler D, Altboum Z. Detection of specific *Bacillus anthracis* spore biomarkers by matrix assisted laser desorption/ionization time-of-flight mass spectrometry. *Rapid Commun Mass Spectrom* 2001;**15**:2110–6.

37. Fagerlund A, Lindback T, Storset AK, Granum PE, Hardy SP. *Bacillus cereus* Nhe is a pore-forming toxin with structural and functional properties similar to the ClyA (HlyE, SheA) family of haemolysins, able to induce osmotic lysis in epithelia. *Microbiology* 2008;**154**(Pt 3):693–704.

38. Fagerlund A, Ween O, Lund T, Hardy SP, Granum PE. Genetic and functional analysis of the cytK family of genes in *Bacillus cereus*. *Microbiology* 2004;**150**(Pt 8):2689–97.

39. Felder KM, Hoelzle K, Wittenbrink MM, Zeder M, Ehricht R, Hoelzle LE. A DNA microarray facilitates the diagnosis of *Bacillus anthracis* in environmental samples. *Lett Appl Microbiol* 2009;**49**:324–31.

40. Feldman S, Pearson TA. Fatal *Bacillus cereus* pneumonia and sepsis in a child with cancer. *Clin Pediatr (Phila)* 1974;**13**:649–51. 654–5.

41. Forsberg LS, Choudhury B, Leoff C, Marston CK, Hoffmaster AR, Saile E, et al. Secondary cell wall polysaccharides from *Bacillus cereus* strains G9241, 03BB87 and 03BB102 causing fatal pneumonia share similar glycosyl structures with the polysaccharides from *Bacillus anthracis*. *Glycobiology* 2011;**21**:934–48.

42. Frankard J, Li R, Taccone F, Struelens MJ, Jacobs F, Kentos A. *Bacillus cereus* pneumonia in a patient with acute lymphoblastic leukemia. *Eur J Clin Microbiol Infect Dis* 2004;**23**:725–8.

43. Fricker M, Ågren J, Segerman B, Knutsson R, Ehling-Schulz M. Evaluation of *Bacillus* strains as model systems for the work on *Bacillus anthracis* spores. *Int J Food Microbiol* 2011;**145**:S129–36.

44. Fricker M, Messelhäußer U, Busch U, Scherer S, Ehling-Schulz M. Diagnostic real-time PCR assays for the detection of emetic *Bacillus cereus* strains in foods and recent food borne outbreaks. *Appl Environ Microbiol* 2007;**73**:1892–8.

45. Funada H, Machi T, Matsuda T. *Bacillus cereus* pneumonia with empyema complicating aplastic anemia–a case report. *Kansenshogaku Zasshi* 1991;**65**:477–80.

46. Gilmore MS, Cruz-Rodz AL, Leimeister-Wachter M, Kreft J, Goebel W. A *Bacillus cereus* cytolytic determinant, cereolysin AB, which comprises the phospholipase C and sphingomyelinase genes: nucleotide sequence and genetic linkage. *J Bacteriol* 1989;**171**:744–53.

47. Gominet M, Slamti L, Gilois N, Rose M, Lereclus D. Oligopeptide permease is required for expression of the *Bacillus thuringiensis plcR* regulon and for virulence. *Mol Microbiol* 2001;**40**:963–75.

48. Gordon RE, Haynes WC, Nor-Nay Pang C. *The genus Bacillus. Agriculture handbook no. 427.* Washington (DC): USDA; 1973.

49. Guillemet E, Tran SL, Cadot C, Rognan D, Lereclus D, Ramarao N. Glucose 6P binds and activates HlyIIR to repress *Bacillus cereus* haemolysin *hlyII* gene expression. *PLoS One* 2013;**8**(2):e55085.

50. Guinebretiere MH, Broussolle V, Nguyen-The C. Enterotoxigenic profiles of food-poisoning and food-borne *Bacillus cereus* strains. *J Clin Microbiol* 2002;**40**:3053–6.

51. Guinebretière MH, Thompson FL, Sorokin A, Normand P, Dawyndt P, Ehling-Schulz M, et al. Ecological diversification in the *Bacillus cereus* group. *Environ Microbiol* 2008; **10**:851–65.

52. Guinebretière MH, Velge P, Couvert O, Carlin F, Debuyser ML, Nguyen-The C. Ability of *Bacillus cereus* group strains to cause food poisoning varies according to phylogenetic affiliation (groups I to VII) rather than species affiliation. *J Clin Microbiol* 2010;**48**:3388–91.

53. Helgason E, Okstad OA, Caugant DA, Johansoen HA, Fouet A, et al. *Bacillus anthracis, Bacillus cereus* and *Bacillus thuringiensis* - one species on the basis of genetic evidence. *Appl Env Microbiol* 2000;**66**:2627–30.

54. Hoffmaster AR, Hill KK, Gee JE, Marston CK, De BK, Popovic T, et al. Characterization of *Bacillus cereus* isolates associated with fatal pneumonias: strains are closely related to *Bacillus anthracis* and harbor *B. anthracis* virulence genes. *J Clin Microbiol* 2006;**44**:3352–60.

55. Jackson SG, Goodbrand RB, Ahmed R, Kasatiya S. *Bacillus cereus* and *Bacillus thuringiensis* isolated in a gastroenteritis outbreak investigation. *Lett Appl Microbiol* 1995;**21**:103–5.

56. Jernigan JA, Stephens DS, Ashford DA, Omenaca C, Topiel MS, et al. Bioterrorism-related inhalational anthrax: the first 10 cases reported in the United States. *Emerg Infect Dis* 2001;**7**:933–44.

57. Jernigan JA, Stephens DS, Ashford DA, Omenaca C, Topiel MS, et al. Bioterrorism-related inhalational anthrax: investigation of bioterrorism-related anthrax, United States, 2001. *Epidemiol Find* 2002;**8**:1019–28.

58. Jevon GP, Dunne Jr WM, Hicks MJ, Langston C. *Bacillus cereus* pneumonia in premature neonates: a report of two cases. *Pediatr Infect Dis J* 1993;**12**:251–3.

59. Jiménez G, Urdiain M, Cifuentes A, López-López A, Blanch AR, Tamames J, et al. Description of *Bacillus toyonensis* sp. nov., a novel species of the *Bacillus cereus* group, and pairwise genome comparisons of the species of the group by means of ANI calculations. *Syst Appl Microbiol* 2013;**36**:383–91.

60. Katsuya H, Takata T, Ishikawa T, Sasaki H, Ishitsuka K, Takamatsu Y, et al. A patient with acute myeloid leukemia who developed fatal pneumonia caused by carbapenem-resistant *Bacillus cereus*. *J Infect Chemother* 2009;**15**:39–41.

61. Koch R. Untersuchungen über Bakterien: V. Die Ätiologie der Milzbrand-Krakheit, begründet auf die Entwicklungsgeschichte des *Bacillus anthracis*. *Cohns Beitr Biol Pflanz* 1876;**2**:277–310.

62. Kolsto AB, Tourasse NJ, Okstad OA. What sets *Bacillus anthracis* apart from other *Bacillus* species? *Annu Rev Microbiol* 2009;**63**:451–76.

63. Kreft J, Berger H, Hartlein M, Muller B, Weidinger G, Goebel W. Cloning and expression in *Escherichia coli* and *Bacillus subtilis* of the hemolysin (cereolysin) determinant from *Bacillus cereus*. *J Bacteriol* 1983;**155**:681–9.

64. Kumar TD, Murali HS, Batra HV. Construction of a non toxic chimeric protein (L1-L2-B) of Haemolysin BL from *Bacillus cereus* and its application in HBL toxin detection. *J Microbiol Methods* 2008;**75**:472–7.

65. Kuroda M, Serizawa M, Okutani A, Sekizuka T, Banno S, Inoue S. Genome-wide single nucleotide polymorphism typing method for identification of *Bacillus anthracis* species and strains among *B. cereus* group species. *J Clin Microbiol* 2010;**48**:2821–9.

66. Leff A, Jacobs R, Gooding V, Hauch J, Conte J, Stulbarg M. *Bacillus cereus* pneumonia. Survival in a patient with cavitary disease treated with gentamicin. *Am Rev Respir Dis* 1977;**115**:151–4.

67. Leski TA, Caswell CC, Pawlowski M, Klinke DJ, Bujnicki JM, Hart SJ, et al. Identification and classification of *bcl* genes and proteins of *Bacillus cereus* group organisms and their application in *Bacillus anthracis* detection and fingerprinting. *Appl Environ Microbiol* 2009;**75**:7163–72.

68. Lund T, De Buyser ML, Granum PE. A new cytotoxin from *Bacillus cereus* that may cause necrotic enteritis. *Mol Microbiol* 2000;**38**:254–61.

69. Martínez-Blanch JF, Sánchez G, Garay E, Aznar R. Development of a real-time PCR assay for detection and quantification of enterotoxigenic members of *Bacillus cereus* group in food samples. *Int J Food Microbiol* 2009;**35**:15–21.

70. Mikkola R, Saris NE, Grigoriev PA, Andersson MA, Salkinoja-Salonen MS. Ionophoretic properties and mitochondrial effects of cereulide: the emetic toxin of *B. cereus*. *Eur J Biochem* 1999;**263**:112–7.

71. Miles G, Bayley H, Cheley S. Properties of *Bacillus cereus* hemolysin II: a heptameric transmembrane pore. *Protein Sci* 2002;**11**:1813–24.

72. Miller JM, Hair JG, Hebert M, Hebert L, Roberts Jr FJ, Weyant RS. Fulminating bacteremia and pneumonia due to *Bacillus cereus*. *J Clin Microbiol* 1997;**35**:504–7.

73. Mock M, Fouet A. Anthrax. *Ann Rev Microbiol* 2001;**55**:647–71.

74. Ngamwongsatit P, Buasri W, Pianariyanon P, Pulsrikarn C, Ohba M, Assavanig A, et al. Broad distribution of enterotoxin genes (*hblCDA*, *nheABC*, *cytK*, and *entFM*) among *Bacillus thuringiensis* and *Bacillus cereus* as shown by novel primers. *Int J Food Microbiol* 2008;**121**:352–6.

75. Oda M, Takahashi M, Matsuno T, Uoo K, Nagahama M, Sakurai J. Hemolysis induced by *Bacillus cereus* sphingomyelinase. *Biochim Biophys Acta* 2010;**1798**:1073–80.

76. Paananen A, Mikkola R, Sareneva T, Matikainen S, Hess M, Andersson M, et al. Inhibition of human natural killer cell activity by cereulide, an emetic toxin from *Bacillus cereus*. *Clin Exp Immunol* 2002;**129**:420–8.

77. Peng H, Ford V, Frampton EW, Restaino L, Shelef LA, Spitz H. Isolation and enumeration of *Bacillus cereus* from foods on a novel chromogenic plating media. *Food Microbiol* 2001;**18**:231–8.

78. Ramarao N, Sanchis V. The pore-forming haemolysins of *Bacillus cereus*: a review. *Toxins* 2013;**5**:1119–39.

79. Rasko DA, Altherr MR, Han CS, Ravel J. Genomics of the *Bacillus cereus* group of organisms. *FEMS Microbiol Lett* 2005;**29**:303–29.

80. Raymond B, Wyres KL, Sheppard SK, Ellis RJ, Bonsall MB. Environmental factor determining the epidemiology and population genetic structure of the *Bacillus cereus* group in the field. *PLoS Pathog* 2010;**6**:e1000905.

81. Ryzhov V, Hathout Y, Fenselau C. Rapid characterization of spores of *Bacillus cereus* group bacteria by matrix-assisted laser desorption-ionization time-of-flight mass spectrometry. *Appl Environ Microbiol* 2000;**66**:3828–34.

82. Sabath LD, Abraham EP. Zinc as a cofactor for cephalosporinase from *Bacillus cereus*. *Biochem J* 1966;**98**:11C–3C.

83. Savini V, Favaro M, Fontana C, Catavitello C, Balbinot A, Talia M, et al. *Bacillus cereus* heteroresistance to carbapenems in a cancer patient. *J Hosp Infect* 2009;**71**:288–90.

84. Savini V, Polilli E, Marrollo R, Astolfi D, Fazii P, D'Antonio D. About the *Bacillus cereus* group. *Intern Med* 2013;**52**:1.

85. Sineva E, Shadrin A, Rodikova EA, Andreeva-Kovalevskaya ZI, Protsenko AS, Mayorov SG, et al. Iron regulates expression of *Bacillus cereus* hemolysin II via global regulator Fur. *J Bacteriol* 2012;**194**:3327–35.

86. Slamti L, Perchat S, Huillet E, Lereclus D. Quorum sensing in *Bacillus thuringiensis* is required for completion of a full infectious cycle in the insect. *Toxins* 2014;**6**: 2239–55.

87. Smith NR, Gordon RE, Clark FE. *Aerobic spore forming bacteria*. Monograph No. 16. Washington (DC): USDA; 1952.

88. Stenfors Arnesen LP, Fagerlund A, Granum PE. From soil to gut: *Bacillus cereus* and its food poisoning toxins. *FEMS Microbiol Rev* 2008;**32**:579–606.

89. Stopler T, Camuescu V, Voiculescu M. Bronchopneumonia with lethal evolution determined by a microorganism of the genus *Bacillus* (*B. cereus*). *Rum Med Rev* 1965;**19**:7–9.

90. Strauss R, Mueller A, Wehler M, Neureiter D, Fischer E, Gramatzki M, et al. Pseudomembranous tracheobronchitis due to *Bacillus cereus*. *Clin Infect Dis* 2001;**33**:E39–41.

91. Sue D, Hoffmaster AR, Popovic T, Wilkins PP. Capsule production in *Bacillus cereus* strains associated with severe pneumonia. *J Clin Microbiol* 2006;**44**:3426–8.

92. Swaminathan B, Gerner-Smidt P, Ng LK, Lukinmaa S, Kam KM, Rolando S, et al. Building PulseNet International: an interconnected system of laboratory networks to facilitate timely public health recognition and response to foodborne disease outbreaks and emerging foodborne diseases. *Foodborne Pathog Dis* 2006;**3**:36–50.

93. Tourasse NJ, Helgason E, Klevan A, Sylvestre P, Moya M, Haustant M, et al. Extended and global phylogenetic view of the *Bacillus cereus* group population by combination of MLST, AFLP, and MLEE genotyping data. *Food Microbiol* 2011;**28**:236–44.

94. Tourasse NJ, Helgason E, Økstad OA, Hegna IK, Kolstø AB. The *Bacillus cereus* group: novel aspects of population structure and genome dynamics. *J Appl Microbiol* 2006;**101**: 579–93.

95. Tourasse NJ, Kolsto AB. Super CAT: a super tree database for combined and integrative multilocus sequence typing analysis of the *Bacillus cereus* group of bacteria (including *B. cereus*, *B. anthracis* and *B. thuringiensis*). *Nucleic Acids Res* 2008;**36**:D461–8.

96. Tourasse NJ, Okstad OA, Kolstø AB. HyperCAT: an extension of the SuperCAT database for global multischeme and multidata type phylogenetic analysis of the *Bacillus cereus* group population. *Database (Oxford)* 2010;**2010**:baq017.

97. Tran SL, Guillemet E, Ngo-Camus M, Clybouw C, Puhar A, Moris A, et al. Haemolysin II is a *Bacillus cereus* virulence factor that induces apoptosis of macrophages. *Cell Microbiol* 2011;**13**:92–108.

98. Tsilia V, Devreese B, de Baenst I, Mesuere B, Rajkovic A, Uyttendaele M, et al. Application of MALDI-TOF mass spectrometry for the detection of enterotoxins produced by pathogenic strains of the *Bacillus cereus* group. *Anal Bioanal Chem* 2012;**404**:1691–702.

99. Uchino Y, Iriyama N, Matsumoto K, Hirabayashi Y, Miura K, Kurita D, et al. A case series of *Bacillus cereus* septicemia in patients with hematological disease. *Intern Med* 2012;**51**:2733–8.

100. Vassileva M, Torii K, Oshimoto M, Okamoto A, Agata N, Yamada K, et al. Phylogenetic analysis of *Bacillus cereus* isolates from severe systemic infections using multilocus sequence typing scheme. *Microbiol Immunol* 2006;**50**:743–9.

101. Walker D. Sverdlovsk revisited: pulmonary pathology of inhalational anthrax versus anthrax-like *Bacillus cereus* pneumonia. *Arch Pathol Lab Med* 2012;**136**:235.

102. Wehrle E, Didier A, Moravek M, Dietrich R, Märtlbauer E. Detection of *Bacillus cereus* with enteropathogenic potential by multiplex real-time PCR based on SYBR Green I. *Mol Cell Probes* 2010;**24**:124–30.

103. Wielinga PR, Hamidjaja RA, Agren J, Knutsson R, Segermanm B, Fricker M, et al. A multiplex real-time PCR for identifying and differentiating *B. anthracis* virulent types. *Int J Food Microbiol* 2011;**145**:S137–44.

104. Wright AM, Beres SB, Consamus EN, Long SW, Flores AR, Barrios R, et al. Rapidly progressive, fatal, inhalation anthrax-like infection in a human: case report, pathogen genome sequencing, pathology, and coordinated response. *Arch Pathol Lab Med* 2011;**135**: 1447–59.

105. Yamada S, Ohashi E, Agata N, Venkateswaran K. Cloning and nucleotide sequence analysis of *gyrB* of *Bacillus cereus, B. thuringiensis, B. mycoides*, and *B. anthracis* and their application to the detection of *B. cereus* in rice. *Appl Environ Microbiol* 1999;**65**:1483–90.

# CHAPTER 7

# *Bacillus cereus* Disease in Children

Daniela Onofrillo

Pediatric Hematology and Oncology Unit, Department of Hematology, Spirito Santo Hospital, Pescara, Italy

## SUMMARY

The role of *Bacillus cereus* as an agent of infections, even serious and lethal ones, is a worrisome concern among the oncologic pediatric population. Children with underlying cancer, including blood malignancies, may develop bacteremia, in particular, sometimes complicated by brain involvement. *Bacillus cereus* may also be responsible for peritonitis in children, related to chronic peritoneal dialysis procedures. Awareness of this organism's relevance as a cause of clinical syndromes outside of food poisoning is warranted, both in adults and in the pediatric hosts, which should lead to efforts targeting a prompt diagnosis and effective, sometimes life-saving, treatment.

## INTRODUCTION

*Bacillus cereus* is an aerobic, Gram-positive, rod-shaped, spore-forming bacterium that is widely distributed in the natural environment and is usually considered a contaminant when recovered from cultures of clinical samples. Although it commonly causes food-borne enteritis, which is mostly benign and self-limiting, it is occasionally responsible for severe infections that are associated with significant morbidity and mortality both in the healthy pediatric population and in high-risk groups of people, such as immunocompromised children, including those receiving treatment for cancer, patients on hemodialysis, and subjects with indwelling venous and cerebrospinal catheters. Being that *B. cereus* is ubiquitous in the environment, it can be isolated from soil, water, dust, air, and other sources, and the gastrointestinal tract and oropharynx of humans can consequently become colonized. The spectrum of described infections other than those involving the gut includes pneumonia, bacteremia, septic shock, cellulitis, panophthalmitis, endocarditis, meningitis, cerebral abscess, and infection-related coagulopathy and hemolysis.[1] The pathogenicity of *B. cereus* is associated with secreted toxins that can induce hemolysis and widespread tissue destruction. The variability in the clinical picture associated with

*The Diverse Faces of Bacillus cereus*
ISBN 978-0-12-801474-5
http://dx.doi.org/10.1016/B978-0-12-801474-5.00007-4

*B. cereus* infection relies on the diversity in its virulence gene patterns, particularly in relation to multiple potential toxin combinations.[2] These compounds are described in a dedicated chapter.[3,4]

## *BACILLUS CEREUS* IN HEALTHY CHILDREN

*Bacillus cereus* is widely recognized as a food-borne pathogen that causes a self-limiting enteritis requiring only symptomatic treatment. The syndrome is mediated by exotoxins, including a diarrheal toxin (enterotoxin) and an emetic toxin (cereulide). However, *B. cereus* food poisoning can be, extremely rarely, fatal in otherwise healthy persons. In 1997 Mahaler et al.[5] reported the case of a 17-year-old boy who died of fulminant liver failure after eating food contaminated with *B. cereus*. Gastrointestinal symptoms developed (even in the patient's father) 30 min after they ate spaghetti with homemade pesto. The father had abdominal pain with diarrhea, but his overall condition remained satisfactory under symptomatic treatment. In contrast, the son had no diarrhea and vomited the initial dose of charcoal despite antiemetic treatment. His condition gradually deteriorated within 2 days, and he became listless. Finally, fulminant hepatic failure, rhabdomyolysis, acute renal impairment, and brain edema were diagnosed, and the patient finally died from liver damage.

In 2005 Dierick et al.[6] reported a case of *B. cereus* food poisoning in five Belgian children after they ate pasta salad. The salad was prepared on a Friday and taken to a picnic on the following day; the remainders had been stored in the fridge until the following Monday evening, when they were served for supper to the children. Six hours after the meal the youngest girl, 7 years of age, started vomiting. She then developed respiratory distress and was taken to the emergency department of a local hospital. Upon arrival, her brothers and sisters started vomiting as well. Because the clinical condition of two children deteriorated rapidly, they were intubated and mechanically ventilated. One of them had severe pulmonary hemorrhage and needed continuous resuscitation. Upon arrival she was moribund, and coma, diffuse bleeding, and severe muscle cramps were documented. She died within 20 min, 13 h after the meal. On autopsy, *B. cereus* was detected in her gut content as well as in the spleen, probably owing to postmortem translocation of the bacterium. A postmortem liver biopsy showed microvascular and extensive coagulation necrosis. Her laboratory values showed severe metabolic acidosis and liver failure. All four of the other children were affected, although to different degrees, fortunately.

The mentioned cases highlight the potential severity of *B. cereus* emetic syndrome along with the importance of adequate refrigeration of prepared food.

*Bacillus cereus* has been associated with a fulminant necrotizing infection resembling gas gangrene in an 8-year-old Australian boy[7] and in fasciitis of the right extremity in a 9-year-old boy living in United States.[8] Both of them developed the infection after a penetrating trauma with a tree branch. These reports further emphasize the relevance and need of recognizing *B. cereus* as a possible cause of severe soft-tissue infection. Hence, it emerges that the organism must reasonably be included in the differential diagnosis of gas gangrene syndromes and necrotizing fasciitis.

Reports dealing with *B. cereus* as a potentially serious agent of liver abscess have been published, as well.[9] For example, Hegarty et al.[10] reported the tale of a healthy 13-year-old boy with a 1-week history of fever, rigors, and a swelling in the right, posterior, lower rib cage. He had been experiencing right-sided abdominal pain for 1 month prior to his request for medical care. On examination, he showed a $12 \times 10$-cm superficial fluctuant mass extending from the right posterior ninth rib to the flank. An ultrasound examination documented a $6 \times 6$-cm abscess in the liver right lower lobe, which had breached the capsule and formed a 6-cm subcutaneous abscess, as confirmed by computed tomography. Fifteen milliliters of purulent material was cultivated and grew *B. cereus*. Of concern, the patient had no underlying immunologic dysfunction.

## *BACILLUS CEREUS* IN ONCOLOGIC CHILDREN

Immunocompromised patients have been known to show an increased risk for serious infections. Although the majority of these conditions are caused by Gram-negative enteric bacteria as well as Gram-positive coccoid organisms, Gram-positive bacilli such as *Bacillus* species have been described as causes of severe infectious scenarios. Whereas most blood *Bacillus* isolates are contaminants, *B. cereus* can be responsible for true bacteremia in cancer children.[11]

Again, the organism may cause primary cutaneous infection in neutropenia patients that are under treatment for cancer or aplastic anemia. In these cases, vesicles or pustules have been described to appear on the limbs, and the reported infections developed in spring or summer; and fortunately they responded to antibiotics. As a take-home message, then, in neutropenia patients, *B. cereus* should be included in the differential diagnosis of isolated vesicles affecting the oncologic pediatric population.[12]

Coming back to *B. cereus* bacteremia, this has been documented in 3% of bacteremic children with cancer; of concern, fulminant *B. cereus* septicemias are frequently lethal and it is estimated that 33–50% of immunocompromised patients with *B. cereus* bloodstream infections develop meningitis or abscess.[13,14]

In a survey by Gaur et al.[14] involving 12 *B. cereus* bacteremic patients with a median age of 11.5 years (range, 4–24 years), three had *B. cereus* even in the cerebrospinal fluid. Two of these children died, while one survived with serious sequelae.

In the patients who died, the time from diagnosis to irreversible damage or death was brief, which emphasizes the need for an increased awareness, an early diagnosis, and an efficacious therapy against *B. cereus*. Risk factors behind a fulminant course or a poor outcome of *B. cereus* bacteremia include an underlying leukemia, in relapse or in the induction phase of chemotherapy; neutropenia; treatment with systemic corticosteroids or third-generation cephalosporins; recent hospitalization; and recent lumbar puncture with intrathecal chemotherapy.

Also, ceftazidime given as monotherapy, which is used frequently as an empiric treatment for neutropenic patients with unexplained fever, would be unlikely to eradicate *B. cereus* bacteremia or to prevent sequelae. Therefore, vancomycin plus a carbapenem would be a good alternative in patients with associated gastrointestinal symptoms or those for which a preliminary report of Gram-positive bacilli in blood cultures exists.

Chou et al.[15] described the case of a 15-year-old girl with an underlying B cell acute lymphoblastic leukemia, who fell into a somnolent state after a 12-h fever, muscle soreness, myalgia in both calves, sore throat, and vomiting. Sepsis due to *B. cereus* was finally identified.

Again, from an analysis of 16 further reported cases[16–20] it emerges that affected children were from 3 to 17 years of age and suffered from acute lymphoblastic leukemia, acute myeloid leukemia, myelodysplastic syndrome, and non-Hodgkin lymphoma or, in general, neutropenia. Gut-related symptoms may be present as well as central nervous system (CNS) lesions. Also, corticosteroid and vancomycin treatment may appear in the anamnesis prior to the sepsis onset.

Six of the 16 patients died, but there were no apparent significant differences in age, chemotherapy, presence of CNS lesions, and use of corticosteroids between survivors and nonsurvivors.

All *B. cereus* isolates were susceptible to vancomycin and imipenem/meropenem but resistant to penicillins and cephalosporins. Also, most strains were susceptible to amikacin, gentamicin, chloramphenicol, and macrolides.

Of concern, Hansford et al.[21] reported the case of an 8-year-old boy undergoing induction therapy for acute lymphoblastic leukemia who developed multifocal *B. cereus* brain abscesses, which highlighted the propensity of this organism to affect the cerebral tissue.

On day 11 of induction he started suffering from generalized headache, epigastric pain, fever, and neutropenia. Abdominal ultrasound imaging documented a moderate free fluid and a thickened and inflamed gall bladder wall with heterogeneous appearance of the pancreatic head. These findings were suggestive of ascending cholangitis. *Bacillus cereus* was identified from blood cultures and vancomycin administered. Cerebral magnetic resonance imaging (MRI) showed multiple peripheral, cortical, and subcortical septic foci within both frontal and parietal lobes, with a dominant right posterior parietal lesion that was observed to cross the parieto-occipital sulcus into the adjacent occipital lobe.

Intravenous ciprofloxacin and meropenem were added. Three days later a new MRI revealed an evolving abscess formation. A presumptive diagnosis of multiple *B. cereus* abscesses was made. Fourteen days after the initial presentation, neutrophil recovery was observed along with improvement in the patient's clinical condition. Ten months after the *B. cereus* infection, the patient was under maintenance chemotherapy, showing good conditions, but had a residual homonymous hemianopia.

Over the past 25 years 11 further cases (involving three children and eight adults) of *B. cereus* brain abscess in cancer patients have been documented. This condition was associated with a high mortality rate (42%) and a significant morbidity. Notably, *B. cereus* bacteremia concomitant with cerebral abscess was associated with induction chemotherapy for acute leukemia in both the child and the adult populations. The gold standard treatment for *B. cereus* brain abscess in immunocompromised patients remains unknown, thus far. The most commonly utilized approach is a prolonged use of systemic antibiotics and sometimes surgery.

Early diagnosis of CNS involvement through neuroimaging is warranted in neutropenic cancer patients with suspected or proven *B. cereus* bacteremia, particularly in those with acute leukemia during induction therapy. Failure to identify such an infection may result in intravenous therapy that is too short, with subsequent risk of relapse and/or significant morbidity.

## *BACILLUS CEREUS* PERITONITIS IN CHILDREN ON CHRONIC PERITONEAL DIALYSIS

Despite technological improvements that have occurred in the field of dialysis connectology and technique, peritonitis remains the most common and most significant complication of peritoneal dialysis (PD) in children. Most children that undergo chronic PD in fact develop no to multiple peritonitis episodes.[22] *Bacillus cereus* infection complicating maintenance PD is extremely rare both in adults and in children and very few cases of peritonitis by this pathogen have been described.[23] However, this organism's role as an opportunistic agent of both local and systemic infections is being increasingly recognized, as widely discussed in this book. Bacteremia has been described to be related to contamination of dialysis equipment, wound or burn infections, and endophthalmitis.

Particularly, also, there have been seven reported cases of *B. cereus* peritonitis involving PD patients as of this writing.[23–28]

The clinical pictures included gastrointestinal symptoms. In three cases, the infection was resolved by virtue of antimicrobial therapy. In four patients, the peritonitis relapsed even though the organism was apparently susceptible to the given treatment and resolved only after PD catheter removal. A possible explanation for this could be the organization of *B. cereus* as biofilm communities on the catheter's inner surface. Antibiotics may not adequately reach this site, in fact, and the infection can then persist unless the PD catheter is removed. In such infections, the PD fluid may be opalescent, with cultures of it growing *B. cereus*.[29] This should not be aprioristically dismissed as a contaminant on such occasions, therefore, but considered as a potential cause of the ongoing pathologic process. Again, the physician must be alert as to the possibility of an unusual bacterial etiology in peritonitis patients on PD who show predominant enteric symptoms. Finally, lack of response of *B. cereus* to a seemingly adequate antibiotic may reflect the need for catheter removal.[26]

## CONCLUDING THOUGHTS

In light of the above discussion, it emerges clearly that, in immunocompromised children or those undergoing dialysis, *B. cereus* isolates should not be indiscriminately regarded as contaminants. In fact, this organism must enter the differential diagnostics of sepsis in patients with immune system impairment, who are also prone to developing *B. cereus* brain involvement. There continues to be a poor outcome among compromised people with meningitis due to *B. cereus*, particularly, with a short interval existing between the disease onset and the establishment of an irreversible injury, which emphasizes the need to explore new tools for early diagnosis and more effective therapy.

# REFERENCES

1. Inoue D, Nagai Y, Mori M, Nagano S, Takiuchi Y, Arima H, et al. Fulminant sepsis caused by *Bacillus cereus* in patients with hematologic malignancies: analysis of its prognosis and risk factors. *Leuk Lymphoma* May 2010;**51**(5):860–9.
2. Horii T, Notake S, Tamai K, et al. *Bacillus cereus* from blood cultures: virulence genes, antimicrobial susceptibility and risk factors for blood stream infection. *FEMS Immunol Med Microbiol* 2011;**63**:202–9.
3. Arnaout MK, Tamburro RF, Bodner SM, et al. *Bacillus cereus* causing fulminant sepsis and hemolysis in two patients with acute leukemia. *J Pediatr Hematol Oncol* 1999;**21**: 431–5.
4. Hirabayashi K, Shiohara M, Saito S, et al. Polymyxin-direct hemoperfusion for sepsis-induced multiple organ failure. *Pediatr Blood Cancer* 2010;**55**:202–5.
5. Mahler H, Pasi A, Kramer JM, Schulte P, Scoging AC, Bär W, et al. Fulminant liver failure in association with the emetic toxin of *Bacillus cereus*. *N Engl J Med* April 17, 1997;**336**(16):1142–8.
6. Dierick K, Van Coillie E, Swiecicka I, Meyfroidt G, Devlieger H, Meulemans A, et al. Fatal family outbreak of *Bacillus cereus*-associated food poisoning. *J Clin Microbiol* August 2005;**43**(8):4277–9.
7. Darbar A, Harris IA, Gosbell IB. Necrotizing infection due to *Bacillus cereus* mimicking gas gangrene following penetrating trauma. *J Orthop Trauma* May–June 2005;**19**(5):353–5.
8. Rosenbaum A, Papaliodis D, Alley M, Lisella J, Flaherty M. *Bacillus cereus* fasciitis: a unique pathogen and clinically challenging sequela of inoculation. *Am J Orthop (Belle Mead NJ)* January 2013;**42**(1):37–9.
9. Latsios G, Petrogiannopoulos C, Hartzoulakis G, Kondili L, Bethimouti K, Zaharof A. Liver abscess due to *Bacillus cereus*: a case report. *Clin Microbiol Infect* December 2003;**9**(12):1234–7.
10. Hegarty RM, Sanka S, Bansal S. Hepatic abscess: presentation in a previously healthy teenager. *Arch Dis Child* February 2013;**98**(2):145.
11. Christenson JC, Byington C, Korgenski EK, Adderson EE, Bruggers C, Adams RH, et al. *Bacillus cereus* infections among oncology patients at a children's hospital. *Am J Infect Control* December 1999;**27**(6):543–6.
12. Henrickson KJ, Shenep JL, Flynn PM, Pui CH. Primary cutaneous *Bacillus cereus* infection in neutropenic children. *Lancet* March 18, 1989;**1**(8638):601–3.
13. Akiyama N, Mitani K, Tanaka Y, et al. Fulminant septicemic syndrome of *Bacillus cereus* in a leukemic patient. *Intern Med* 1997;**36**:221–6.
14. Gaur AH, Patrick CC, McCullers JA, Flynn PM, Pearson TA, Razzouk BI, et al. *Bacillus cereus* bacteremia and meningitis in immunocompromised children. *Clin Infect Dis* May 15, 2001;**32**(10):1456–62.
15. Chou YL, Cheng SN, Hsieh KH, Wang CC, Chen SJ, Lo WT. *Bacillus cereus* septicemia in a patient with acute lymphoblastic leukemia: a case report and review of the literature. *J Microbiol Immunol Infect* August 5, 2013. pii: S1684-1182(13)00114-X. http://dx.doi.org/10.1016/j.jmii.2013.06.010. [Epub ahead of print].
16. Feldman S, Pearson TA. Fatal *Bacillus cereus* pneumonia and sepsis in a child with cancer. *Clin Pediatr (Phila)* 1974;**13**:649–51.
17. Yoshida H, Moriyama Y, Tatekawa T, Tominaga N, Teshima H, Hiraoka A, et al. Two cases of acute myelogenous leukemia with *Bacillus cereus* bacteremia resulting in fatal intracranial hemorrhage. *Rinsho Ketsueki* 1993;**34**:1568–72.
18. Musa MO, Al Douri M, Khan S, Shafi T, Al Humaidh A, Al Rasheed AM. Fulminant septicaemic syndrome of *Bacillus cereus*: three case reports. *J Infect* 1999;**39**:154–6.
19. Jenson HB, Levy SR, Duncan C, McIntosh S. Treatment of multiple brain abscess caused by *Bacillus cereus*. *Pediatr Infect Dis J* 1989;**8**:795–8.

20. Nishikawa T, Okamoto Y, Tanabe T, Kodama Y, Shinkoda Y, Kawano Y. Critical illness polyneuropathy after *Bacillus cereus* sepsis in acute lymphoblastic leukemia. *Intern Med* 2009;**48**:1175–7.

21. Hansford JR, Phillips M, Cole C, Francis J, Blyth CC, Gottardo NG. *Bacillus cereus* bacteremia and multiple brain abscesses during acute lymphoblastic leukemia induction therapy. *J Pediatr Hematol Oncol* April 2014;**36**(3):e197–201.

22. Bakkaloglu SA, Warady BA. Difficult peritonitis cases in children undergoing chronic peritoneal dialysis: relapsing, repeat, recurrent and zoonotic episodes. *Pediatr Nephrol* September 18, 2014;**30**(9):1397–406. [Epub ahead of print].

23. Ruiz SR, Reyes GM, Campos CT, Jimenez VL, Rojas RT, de la Fuente CG, et al. Relapsing *Bacillus cereus* peritonitis during automated peritoneal dialysis. *Perit Dial Int* September–October 2006;**26**(5):612–3.

24. Balakrishnan I, Baillod RA, Kibbler CC, Gillespie SH. *Bacillus cereus* peritonitis in a patient being treated with continuous ambulatory peritoneal dialysis. *Nephrol Dial Transpl* 1997;**12**:2447–8.

25. Al-Wali W, Baillod R, Hamilton-Miller JM, Brumfitt W. Detective work in continuous ambulatory peritoneal dialysis. *J Infect* 1990;**20**:151–4.

26. Al-Hilali N, Nampoory MRN, Johny KV, Chugh TD. *Bacillus cereus* peritonitis in a chronic peritoneal dialysis patient. *Perit Dial Int* 1997;**17**:514–5.

27. Pinedo S, Bos AJ, Siegert CEH. Relapsing *Bacillus cereus* peritonitis in two patients on peritoneal dialysis. *Perit Dial Int* 2002;**22**:424–6.

28. Biasoli S, Chiaramonte S, Fabris A, Feriani M, Pisani F, et al. *Bacillus cereus* as agent of peritonitis during peritoneal. *Perit Dial Int* November–December 2006;**26**(6):715–6.

29. Monteverde ML, Sojo ET, Grosman M, Hernandez C, Delgado N. Relapsing *Bacillus cereus* peritonitis in a pediatric patient on chronic peritoneal dialysis. *Perit Dial Int* September–October 1997;**17**(5):514–5.

# CHAPTER 8

# *Bacillus cereus* Disease Other Than Food-Borne Poisoning

**Giovanni Gherardi**
Department of Biomedical Sciences, Campus Bio-Medico University of Rome, Rome, Italy

## SUMMARY

In addition to food poisoning and airway infections, which are discussed in dedicated chapters, *Bacillus cereus* has been known to cause a number of systemic and local infections, through the years, involving both immunologically compromised and immunocompetent individuals. Among those most commonly infected, we find neonates, intravenous drug abusers, patients sustaining traumatic or surgical wounds, and those having indwelling catheters. The spectrum of diseases includes fulminant bacteremia, central nervous system (CNS) involvement (with meningitis and brain abscesses), endophthalmitis, pneumonia, and gas gangrene-like cutaneous infections, just to name a few.

## NOSOCOMIAL INFECTIONS

By virtue of the ubiquitous distribution of *Bacillus* spores in the environment as well as in hospitals, *B. cereus* is usually dismissed as a contaminant when isolated from clinical materials, and outbreaks of nosocomial pseudo-infections have been well documented, especially pseudo-bacteremias.[1–3] Pseudo-outbreaks of pneumonia and bloodstream infections have even been traced back to contaminated fiber-optic bronchoscopy equipment[4,5] and ethyl alcohol,[2] respectively.

The term pseudo-outbreak means a situation in which a microorganism is found in culture at a greater rate than expected and that cannot be correlated, from a clinical perspective, with the supposed infection implied by results of cultures.[6]

There is, however, no doubt that *Bacillus* spores (in general) permeate hospital environments, thus contributing to nosocomial infections, and *Bacillus* species, in fact, have gained increasing notoriety as agents of

*The Diverse Faces of Bacillus cereus*
ISBN 978-0-12-801474-5
http://dx.doi.org/10.1016/B978-0-12-801474-5.00008-6

93

nosocomial outbreaks in immunocompromised hospitalized hosts.[7,8] Environmental sources that have been identified for this species include specifically contaminated linens,[9–11] gloves,[3] hands of operators,[12] air-filtration and ventilation along with fiber-optic bronchoscopy equipment,[4,5,13] intravenous devices,[14] alcohol-based hand-wash solutions,[2] tubes for specimen collection, balloons for manual ventilation,[15] and reused towels.[16]

As the vast majority of *Bacillus* species (with the exception of *Bacillus anthracis*) cultivated from blood and even from open wounds are usually labeled as contaminants, it is crucial for the hospital microbiology laboratory to alert infection control practitioners if a sudden increase in the recovery of *B. cereus* emerges. In this case, in fact, *B. cereus* isolates should be sent to reference centers for serotyping and/or genotypic fingerprinting,[17] to establish whether isolates are genotypically related.

Concerning catheter-related *B. cereus* septicemias, it has been shown that the organism can produce biofilms,[18,19] which can allow it to attach to catheters. *Bacillus cereus* isolates associated with nosocomial bloodstream infections, in fact, are known to form microbial communities on the device surface, thus resulting in persistent infection unless the device is removed. Of concern, the release of so-called planktonic cells from the biofilm can, in turn, lead to formation of additional biofilms.[20] The crucial issue is that while antibiotics may affect planktonic invasion, sessile organisms are spared, resulting in recurrent or persistent bacteremia. Biofilm formation on inert surfaces may also contribute to *B. cereus* persistence in the nosocomial environment in addition to the spore survival.

## OCULAR INFECTION

*Bacillus cereus* is an important ocular pathogen and can cause a rapidly progressive, refractory-to-treatment endophthalmitis. Cases of ocular infections are quite infrequent and usually follow penetrating trauma of the eye. Moreover, eye infections may occur by virtue of hematogenous spread or, occasionally, eye surgery. The dramatic, severe clinical pictures observed in such syndromes, when caused by *B. cereus*, are often associated with significant vision loss and loss of the eye itself, which can occur within 24–48 h from the disease onset.[21] In general, endophthalmitis is a vision-threatening eye disease resulting from traumatic or systemic microbial infection of the interior of the eye,[22] with the outcome varying depending on the microbial agent involved, the rapidity of treatment, and the host response. The typical ophthalmic lesion is represented by a corneal ring abscess accompanied by rapid progression of pain, chemosis, proptosis, retinal hemorrhage, and perivasculitis. Systemic manifestations can also occur, which include fever, leukocytosis, and general malaise.[23]

Endophthalmitis caused by *B. cereus*, in particular, is characterized by a devastating malignant eye infection with a rapid disease progression and is known to be associated with production of several extracellular tissue-destructive virulence factors released by the organism.[24–26] During the first half of the twentieth century, Gram-positive bacilli isolated from cases of endophthalmitis were not identified to the species level and were all generally grouped as *Bacillus subtilis*.[27] In one case of endophthalmitis, a clinical isolate was successfully identified as *B. cereus*[27] following the identification criteria described earlier in *Bergey's Manual of Determinative Bacteriology*.[28] As the patient's clinical presentation closely mimicked those of earlier reports of endophthalmitis attributed to *B. subtilis*, Davenport and Smith hypothesized that those earlier reports were to be reasonably attributed to *B. cereus*, rather than *B. subtilis*.

*Bacillus cereus* endophthalmitis can generally be classified into two categories based on the source of the infection: exogenous, which is the most common and is attributable to penetrating trauma of the eye, and endogenous, meaning subsequent to hematogenous spread of the microorganism from a distant site or through direct intravenous acquisition, such as in cases of blood transfusion,[29] indwelling devices, contaminated needles, or injection drug receipt.[30–33] Moreover, cases of ocular infections acquired by iatrogenic administration of medications such as B vitamins[34] and insulin[12] have been described.[34]

Eye infections following hematogenous spread of *B. cereus* have been, indeed, described among intravenous drug (i.e., heroin) abusers and are then attributable to contaminated drugs[31,35] or injection equipment.[36]

*Bacillus cereus* keratitis and other more severe ocular infections associated with contact lens wear have been reported, too.[37] Eye infections caused by *B. cereus* and involving contact lens wearers have been known to be associated with the acquisition of the microorganism from contaminated contact lens care systems.[38] It has been pointed out particularly that *Bacillus* spores can both survive multiple heat disinfection treatments and be resistant to chemical disinfection procedures used for the commonly recommended lens care systems.[38]

Indeed, suspicion of the presence of *B. cereus* in eye infection following penetrating ocular trauma has to be considered in those cases related to occupation, e.g., metalworkers,[23] and, particularly, a history of living or working in a rural area or an agricultural setting.[39]

Based on previous reports of *B. cereus* nosocomial infections, contaminated fomites such as gauze, linens, and ventilators, in addition to health-care workers' hands, may have served as the source of *B. cereus* outbreaks. Also, an ocular infection by *B. cereus* should be suspected after cataract extraction surgery.[40]

Rapid diagnosis of endogenous and exogenous *B. cereus* endophthalmitis should be attempted by immediate anterior-chamber paracentesis and, if no growth of the microorganism is observed, a second vitreous aspiration after a short interval should be carried out for microbiological evaluation,[31] along with collection of blood samples for cultures.[41]

Because of the rapid tissue destruction observed with severe *B. cereus* eye infections, in particular in those cases following penetrating trauma with a soil-contaminated foreign body, rapid therapeutic intervention is strongly mandatory irrespective of the results of microbiological tests.[39] The efficacy of intravitreal antibiotics, namely, 1000 μg of vancomycin in combination with 400 μg of amikacin, has been suggested as a successful therapeutic choice.[42] Several factors are reported to contribute to the outcome of *B. cereus* endophthalmitis, such as rapidity and correctness of treatment after injury and the condition of the eye upon the patient's presentation.[43,44] Systemic antibiotics have been also used in combination with local intravitreal antimicrobials, but it should be emphasized that vancomycin and aminoglycosides do not readily penetrate into the vitreous fluid[45] owing to the protective effect of the blood–ocular fluid barrier; this makes intravitreal antibiotic administration the most effective way to treat these infections.[44]

In naturally acquired *B. cereus* infections and in cases of experimentally induced endophthalmitis it has been reported that ocular entrance of the bacterium results in a massive and quick destruction of the eye, usually within 12–18 h[43] and, in many instances, vision loss occurs regardless of the therapeutic and surgical intervention. This occurs especially owing to delayed antibiotic treatment, the effects of toxin production by the infecting strain, and migration and sequestration of the organism (*Bacillus* is motile) out of antibiotic reach.[46] It is well established, in fact, that *B. cereus* produces a panel of tissue-destructive exotoxins that contribute to the devastating outcomes observed in endophthalmitis cases.[24] Also, it has been suggested that a poor outcome may be observed even after adequate antibiotic treatment owing to the persistence of a tissue-destructive activity regardless of the antibiotic bacterial killing.[24] In experimental rabbit models, among the exotoxins incriminated in destructive endophthalmitis,[24,47] hemolysin BL (a tripartite dermonecrotic vascular permeability factor), a crude exotoxin derived from cell-free *B. cereus* culture filtrates, phosphatidylcholine-preferring phospholipase C (PC-PLC), and collagenases represent the most important ones. All together, these factors seem to stand behind the retinal toxicity, necrosis, and blindness in experimentally infected rabbit eyes. In particular, the reported toxicity of PC-PLC seemed to represent a direct result of the propensity of the secreted enzyme to affect phospholipids that are abundant in retinal tissue.[48] Evidence seems, however, to demonstrate that intraocular inflammation and retinal toxicity may occur irrespective of the presence of hemolysin BL, implying the role of other pathogenetic factors.[49]

In an experimental rabbit model used to study the pathogenesis of bacterial endophthalmitis caused by the Gram-positive ocular pathogens *Staphylococcus aureus*, *Enterococcus faecalis*, and *B. cereus*, it was shown that endophthalmitis caused by *B. cereus* had a more rapid and virulent course than what was observed with the other two bacterial species.[22] Additionally, *B. cereus* intraocular growth was significantly greater than that observed for *S. aureus* and *E. faecalis*. Analysis of the bacterial location within the eye showed that the motile *B. cereus* rapidly migrates from posterior to anterior segments during infection[46]; it has in fact been shown that wild-type motile and nonmotile *B. cereus* strains grow equally in the vitreous fluid, but the motile swarming strains migrate to the anterior segment during infection, thus inducing a more severe anterior segment disease than the nonswarming strains.

Bacterial swarming is a specialized form of surface translocation undertaken by flagellated species. Swarm cell population typically undergoes morphological differentiation from short bacillary forms to filamentous, multinucleate, and hyperflagellated swarm cells with nucleoids evenly distributed along the lengths of the filaments.[50–52] Although the differentiated cells do not replicate, they can rapidly migrate away from the colony in organized groups, which comprise the advancing rim of growing colonies.[50,52,53] When the swarming motility collectively stops, swarm cells differentiate back into the short bacillary forms. Swarming is thought to represent an important mechanism by which flagellated microorganisms are able to traverse environmental niches allowing colonization of host mucosal surfaces.[54] Moreover, swarming can play a role in host–pathogen interactions by leading to an increase in the production of specific virulence factors.[51,54] Regarding *B. cereus*, a correlation between swarming and hemolysin BL secretion has been reported.[51] Swarming cells have been observed in fact to produce the highest levels of toxin, thus suggesting that swarming *B. cereus* strains may have a higher virulence potential than nonswarming ones.

The eye is protected from inflammatory cells and blood constituents by the blood–ocular barrier systems that separate the interior portion from the bloodstream and maintain the transparency and function of the interior compartment.[55] Two main blood–ocular barriers are present, which are the blood–aqueous barrier and the blood–retinal barrier, with tight junctions that hamper the free diffusion of molecules.[48,56] In experimental *B. cereus* endophthalmitis[55] it was documented that the bacterium increases the permeability of the blood–retinal barrier as early as 4 h postinfection by disrupting the tight junctions between the endothelial cells and the basement membrane of retinal capillaries and retinal pericytes. Such changes in the blood–ocular barrier contribute to the loss of retinal structure and function.[55,57]

## CENTRAL NERVOUS SYSTEM INFECTIONS

CNS infections represent another important, although infrequent, type of syndrome caused by *B. cereus*. CNS invasion by *B. cereus* leads to meningitis,[58,59] meningoencephalitis,[60] subarachnoid hemorrhage,[61–63] and brain abscesses,[64–66] with these clinical conditions involving pediatric[67,68] and adult immunocompromised patients,[65] mostly those with underlying neoplastic diseases and immunosuppression (i.e., subjects with leukemia or other malignancies). CNS infections generally occur secondarily to *B. cereus* bacteremia or following induction chemotherapy.[60,63,67,69,70]

The pathogenesis of *B. cereus* CNS infection is in most cases obscure, although several risk factors are worthy of consideration. A substantial number of patients developed necrotizing brain lesions following intrathecal induction chemotherapy, a procedure that, in addition to inducing neutropenia, could introduce ubiquitous *B. cereus* spores if environmental sources and fomites are contaminated.[60,63,67,70] Other routes of acquisition include bacteremia from a distal site and infected central venous catheters.[67] Several authors have advanced the possibility that the gastrointestinal tract is a potential source of *B. cereus* in CNS infections, because gastrointestinal symptoms (nausea, vomiting, epigastric pain, diarrhea) suggestive of food poisoning have been documented prior to CNS disease onset or concomitant with it.[60–62,67] This means that *B. cereus* is acquired from an exogenous source, e.g., food or water, leading to gastric invasion, mucosal necrosis, and spread to the liver and CNS.[60–62,67,71–75]

Two published unusual cases of *B. cereus* meningitis in leukemic hosts are of concern.[76,77] In the first, a 19-year-old patient was colonized with *B. cereus* in the oropharyngeal tract (the patient died within 48 h of the meningitis onset). Notably, the food fed to the patient contained a small quantity of *B. cereus*. The second case[77] involved an 18-year-old patient who, during his last day of treatment prior to protective isolation, was allowed to take a short walk on the hospital grounds. By accident, the patient fell, incurring a minor abrasion on his forearm; it was washed and cleansed and regarded as unimportant but, 2 days later, the wound became infected with *B. cereus*, which was also collected from the blood, and a computed-tomography scan revealed signs of a damaged blood–brain barrier potentially compatible with toxic encephalitis. Three days after the initial injury, the patient died from respiratory arrest.

## WOUND AND CUTANEOUS INFECTIONS

Wound infections (sometimes necrotic and gangrenous), usually open factures, caused by *B. cereus* have been reported mostly in healthy people, following surgery, road traffic and other accidents, scalds, burns, plaster fixation, drug injection, and close-range gunshot and nail-bomb injuries.[78,79] The similarity between *B. cereus* wound infections and clostridial myonecrosis or streptococcal necrotizing fasciitis has to be considered when a patient complains of either of these kinds of suspected soft-tissue infections. The mimicry of *B. cereus* myonecrosis compared to that caused by *Clostridium perfringens* may in part be attributed to the pathogenic activity of phospholipase, which in both bacterial species is responsible for increased capillary permeability, platelet aggregation, hemolysis, and myonecrosis.[80–82] Indeed, mixed *B. cereus* and *C. perfringens* infections may occur following traumatic accidents involving soil and water contamination. Therefore, a rapid and accurate microbiological diagnosis is critical and essential for an appropriate antibiotic and surgical treatment, to avoid complications.[80–82]

*Bacillus cereus* infections following bodily trauma, by penetrating objects or as a consequence of burns or motor vehicle-related accidents, have been also reported. Depending on the environment in which the trauma takes place, different microbial species can be involved in traumatic cutaneous infections, with the predominant organisms being *Pseudomonas aeruginosa*, *Aeromonas* spp., and *Vibrio* spp.[83] *Bacillus* species isolated from trauma-induced wounds are usually regarded as contaminants until a more dramatic complication occurs, such as sepsis or necrotizing fasciitis. In more recent years, the recognition of *B. cereus* as a pathogen infecting individuals who suffer from traumatic injuries has been documented.[79,83–86] Spores produced can be disseminated into the environment through dust, water, and food. Wound contamination with *B. cereus* spores can take place at the time of initial trauma because of their ubiquitous presence in the environment. Thus, the clinical consequences of this widespread dispersion of spores are represented by open-wound postsurgical or posttraumatic infections[84] or infections of lesions. following close-range gunshot and nail-bomb injuries,[86] injection-drug abuse,[85] ground-contracted open-wound fractures,[35] and severe war wounds.[87] Moreover, the ability of *B. cereus* spores to persist in plaster-impregnated gauze,[88] incontinence pads, and hospital linens,[10,11] as well as in many antiseptics such as chlorhexidine, povidone iodine, and alcohol,[2,79] has been described, and this allows the bacterium to function as a secondary nosocomial invader of sites of traumatic injury.

Primary cutaneous infections may occur in both immunosuppressed[81,89–92] and nonimmunosuppressed individuals, with the latter being usually involved in traumatic incidents.[79,84,85,87] *Bacillus cereus* spores, which are widespread in soil, may be common in the hospital environment; subsequently, spores of *B. cereus* may enter the skin of hands and feet that are in contact with the environment through skin abrasions, as described for *B. anthracis*.[89,92] Similar to *B. anthracis* infections, the evolving skin lesions caused by *B. cereus* start with a papule, which then becomes serous or serosanguinous and develops a black eschar.[91] The elaboration of the various *B. cereus* exotoxins, including dermonecrotic toxin, may well account for eschar production in *B. cereus* cutaneous infections.[93]

## ENDOCARDITIS

*Bacillus cereus* endocarditis in drug addicts and in patients with an intravascular device has been also reported. In a review by Steen et al.,[94] the rates of morbidity and mortality associated with *B. cereus* endocarditis have been found to be high among patients with valvular heart disease.[95–98] In addition, several cases of *B. cereus* endocarditis have been observed among patients with pacemakers,[98,99] prosthetic mitral valves,[95,96] and other underlying conditions.[100,101]

With the exception of intravenous drug abuse-associated *B. cereus* endocarditis, the source of the microorganism in many cases is somewhat unclear. In patients with an inserted pacemaker or prosthetic valve in place, it could be hypothesized that asymptomatic *B. cereus* bacteremia could induce endocarditis. Alternatively, the repeated use of a venipuncture site for heparin injections could be the venue for introducing *B. cereus* into the blood-stream.[96] *Bacillus cereus* invasion of the gastrointestinal tract under appropriate conditions, as outlined above, and through hematological spread and colonization of a susceptible cardiac valve could represent another conceivable cause of *B. cereus* endocarditis.

## BONE INFECTION

Osteomyelitis caused by *B. cereus* has been reported, although infrequently.[36,98,102,103] *Bacillus cereus* osteomyelitis has been reported in association with intravenous drug addiction and surgical trauma and in individuals who had sustained a motor vehicle accident.[35] Bone infections due to *B. cereus* have been found to be monomicrobial or mixed with other pathogens, such as *S. aureus*.[102]

## URINARY TRACT INFECTIONS

A case of urinary tract infection caused by *B. cereus* has been documented in a 71-year-old woman with invasive bladder cancer.[104] The patient underwent radical cystectomy and percutaneous left ureterostomy, after which an indwelling urethral catheter was placed and cephalosporin and other antibiotics were administered. Five weeks after surgery, the patient complained of fever, shaking chills, and pyuria, and pyelonephritis of the left kidney was diagnosed. A urine culture grew *B. cereus*, and the authors supposed that the organism, first colonizing the catheter, ascended into the urinary system through ureterostomy irrigation. Among the virulence mechanisms behind *B. cereus* pathogenicity, adherence to catheters via biofilm formation and motility could have easily accounted for the ascension of the bacterium into the urinary tract. Occult listings of *B. cereus* urinary tract infections have also cited instrumentation as a prelude to infection.[36]

## REFERENCES

1. Berger SA. Pseudobacteremia due to contaminated alcohol swabs. *J Clin Microbiol* 1983;**18**:974–5.
2. Hsueh P-R, Teng LJ, Yang PC, Pan HL, Ho SW, Luh KT. Nosocomial pseudobacteremia caused by *Bacillus cereus* traced to contaminated ethyl alcohol from a liquor factory. *J Clin Microbiol* 1999;**37**:2280–4.
3. York MK. *Bacillus* species pseudobacteremia traced to contaminated gloves used in collection of blood from patients with acquired immunodeficiency syndrome. *J Clin Microbiol* 1990;**28**:2114–6.

4. Goldstein B, Abrutyn E. Pseudo-outbreak of *Bacillus* species related to fiberoptic bronchoscopy. *J Hosp Infect* 1985;**6**:194–200.
5. Richardson AJ, Rothburn MM, Roberts C. Pseudo-outbreak of *Bacillus* species: related to fibreoptic bronchoscopy. *J Hosp Infect* 1986;**7**:208–10.
6. Maki DG. Through a glass darkly: nosocomial pseudoepidemics and pseudobacteremias. *Arch Intern Med* 1980;**140**:26–8.
7. Ozkocaman V, Ozcelik T, Ali R, Ozkalemkas F, Ozkan A, Ozakin C, et al. *Bacillus* spp. among hospitalized patients with haematological malignancies: clinical features, epidemics and outcomes. *J Hosp Infect* 2006;**64**:169–76.
8. Richards V, van der Auwera P, Snoeck R, Daneau D, Meunier F. Nosocomial bacteremias caused by *Bacillus* species. *Eur J Clin Microbiol Infect Dis* 1988;**7**:783–5.
9. Avashia SB, Wiggens WS, Lindley C, Hoffmaster AH, Drumgoole R, Nekomoto T, et al. Fatal pneumonia among metalworkers due to inhalation exposure to *Bacillus cereus* containing *Bacillus anthracis* toxin genes. *Clin Infect Dis* 2007;**44**:414–6.
10. Barrie D, Wilson JA, Hoffman PN, Kramer JM. *Bacillus cereus* meningitis in two neurological patients: an investigation into the source of the organism. *J Infect* 1992;**25**:291–7.
11. Barrie D, Hoffman PN, Wilson JA, Kramer JM. Contamination of hospital linen by *Bacillus cereus*. *Epidemiol Infect* 1994;**113**:297–306.
12. Motoi N, Ishida T, Nakano I, Akiyama N, Mitani K, Hirai H, et al. Necrotizing *Bacillus cereus* infection of the meninges without inflammatory reaction in a patient with acute myelogenous leukemia: a case report. *Acta Neuropathol* 1997;**93**:301–5.
13. Bryce EA, Smith JA, Tweeddale M, Andruschak BJ, Maxwell MR. Dissemination of *Bacillus cereus* in an intensive care unit. *Infect Control Hosp Epidemiol* 1993;**14**:459–62.
14. Hernaiz C, Picardo A, Alos JI, Gomez-Garces JL. Nosocomial bacteremia and catheter infection by *Bacillus cereus* in an immunocompetent patient. *Clin Microbiol Infect* 2003;**9**:973–5.
15. Van Der Zwet WC, Parlevliet GA, Savelkoul PH, Stoof J, Kaiser PM, Van Furth AM, et al. Outbreak of *Bacillus cereus* infections in a neonatal intensive care unit traced to balloons used in manual ventilation. *J Clin Microbiol* 2000;**38**:431–6.
16. Dohmae S, Okubo T, Higuchi W, Takano T, Isobe H, Baranovich T, et al. *Bacillus cereus* nosocomial infection from reused towels in Japan. *J Hosp Infect* 2008;**69**:361–7.
17. Liu PY, Ke O, Chen S. Use of pulsed-field gel electrophoresis to investigate a pseudo-outbreak of *Bacillus cereus* in a pediatric unit. *J Clin Microbiol* 1997;**35**:1533–5.
18. Auger S, Ramarao N, Faille C, Fouet A, Aymerich S, Gohar M. Biofilm formation and cell surface properties among pathogenic and nonpathogenic strains of the *Bacillus cereus* group. *Appl Environ Microbiol* 2009;**75**:6616–8.
19. Kuroki R, Kawakami K, Qin L, Kaji C, Watanabe K, Kimura Y, et al. Nosocomial bacteremia caused by biofilm-forming *Bacillus cereus* and *Bacillus thuringiensis*. *Intern Med* 2009;**48**:791–6.
20. Costerton JW, Stewart PS, Greenberg EP. Bacterial biofilms: a common cause of persistent infection. *Science* 1999;**284**:1318–22.
21. Miller JJ, Scott IU, Flynn HW, Smiddy WE, Murray TG, Berrocal A, et al. Endophthalmitis caused by *Bacillus* species. *Am J Ophthalmol* 2008;**145**:883–8.
22. Callegan MC, Booth MC, Jett BD, Gilmore MS. Pathogenesis of Gram-positive bacterial endophthalmitis. *Infect Immun* 1999;**67**:3348–56.
23. Martinez MF, Haines T, Waller M, Tingey D, Gomez W. Probable occupational endophthalmitis. *Arch Environ Occup Health* 2007;**62**:157–60.
24. Beecher DJ, Pulido JS, Barney NP, Wong ACL. Extracellular virulence factors in *Bacillus cereus* endophthalmitis: methods and implication of involvement of hemolysin BL. *Infect Immun* 1995;**63**:632–9.
25. Drobniewski FA. *Bacillus cereus* and related species. *Clin Microbiol Rev* 1993;**6**:324–38.

26. Callegan MC, Cochran DC, Ramadan RT, Chodosh J, McLean C, Stroman DW. Virulence factor profiles and antimicrobial susceptibilities of ocular *Bacillus* isolates. *Curr Eye Res* 2006;**31**:693–702.

27. Davenport R, Smith C. Panophthalmitis due to an organism of the *Bacillus subtilis* group. *Br J Ophthalmol* 1952;**36**:389–92.

28. Holt JG. *Bergey's manual of determinative biology*. Baltimore (MD): Lippincott Williams & Wilkins; 1994.

29. Kerkenezov N. Panophthalmitis after a blood transfusion. *Br J Ophthalmol* 1953;**37**:632–6.

30. Grossniklaus H, Bruner H, Frank WE, Purnell EW. *Bacillus cereus* panophthalmitis appearing as an acute glaucoma in a drug addict. *Am J Ophthalmol* 1985;**100**:334.

31. Masi RJ. Endogenous endophthalmitis associated with *Bacillus cereus* bacteremia in a cocaine addict. *Ann Ophthalmol* 1978;**10**:1367–70.

32. Shamsuddin D, Tuazon CV, Levy C, Curtin J. *Bacillus cereus* panophthalmitis: source of the organism. *Rev Infect Dis* 1982;**4**:97–103.

33. Tuazon CV, Hill R, Sheagren JN. Microbiologic study of street heroin and injection paraphernalia. *J Infect Dis* 1974;**129**:327–9.

34. Bouza E, Grant S, Jordan MC, Yook RW, Sulit HL. *Bacillus cereus* endogenous panophthalmitis. *Arch Ophthalmol* 1979;**97**:498–9.

35. Wong MT, Dolan MJ. Significant infections due to *Bacillus* species following abrasions associated with motor vehicle-related trauma. *Clin Infect Dis* 1992;**15**:855–7.

36. Tuazon CV, Murray HW, Levy C, Solny MN, Curtin JA, Sheagren JN. Serious infections from *Bacillus* species. *JAMA* 1979;**241**:1137–40.

37. Pinna A, Sechi LA, Zanetti S, Esai D, Delogu G, Cappuccinelli P, et al. *Bacillus cereus* keratitis associated with contact lens wear. *Ophthalmology* 2001;**108**:1830–4.

38. Donzis PB, Mondino BJ, Weisman BA. *Bacillus* keratitis with contaminated contact lens case system. *Am J Ophthalmol* 1988;**105**:195–7.

39. David RB, Kkirkby GR, Noble BA. *Bacillus cereus* endophthalmitis. *Br J Ophthalmol* 1994;**78**:577–80.

40. Simini B. Outbreak of *Bacillus cereus* endophthalmitis in Rome. *Lancet* 1998;**351**:1258.

41. Greenwald MJ, Wohl LG, Sell CH. Metastatic bacterial endophthalmitis: a contempory reappraisal. *Surv Ophthalmol* 1986;**31**:81–101.

42. Vahey JB, Flynn HW. Results in the management of *Bacillus* endophthalmitis. *Ophthalmic Surg* 1991;**22**:681–6.

43. Callegan MC, Engelbert M, Parke II DW, Jett BD, Gilmore MS. Bacterial endophthalmitis: epidemiology, therapeutics, and bacterium-host interactions. *Clin Microbiol Rev* 2002;**15**:111–24.

44. Foster JR, Martinez JA, Murray TG, Rubsamen PE, Flynn HW, Foster RK. Useful visual outcomes after treatment of *Bacillus cereus* endophthalmitis. *Ophthalmology* 1996;**103**:390–7.

45. Ferencz JR, Assia EI, Diamantstein L, Rubinstein E. Vancomycin concentration in the vitreous after intravitreal administration for post operative endophthalmitis. *Arch Ophthalmol* 1999;**117**:1023–7.

46. Callegan MC, Novasad BD, Ramirez R, Ghelardi G, Senesi S. Role of swarming migration in the pathogenesis of *Bacillus* endophthalmitis. *Invest Ophthalmol Vis Sci* 2006;**47**:4461–7.

47. Beecher DJ, Olsen TW, Somers EB, Wong ACL. Evidence for contribution of tripartite hemolysin BL, phosphatidylcholine-preferring phospholipase C, and collegenase to virulence of *Bacillus cereus* endophthalmitis. *Infect Immun* 2000;**68**:5269–76.

48. Berman ER. *Biochemistry of the eye*. New York (NY): Plenum Press; 1991.

49. Callegan MC, Jett BD, Hancock LE, Gilmore MS. Role of hemolysin BL in the pathogenesis of extraintestinal *Bacillus cereus* infections as assessed in an endophthalmitis model. *Infect Immun* 1999;**67**:3357–66.

50. Eberl L, Christiansen G, Molin S, Givskov M. Differentiation of *Serratia liquefaciens* into swarm cells is controlled by expression of the *flhD* master operon. *J Bacteriol* 1996; **178**:554–9.

51. Ghelardi E, Celandroni F, Salvetti S, Ceragoli M, Beecher DJ, Senesi S, et al. Swarming behavior of and hemolysin secretion by *Bacillus cereus*. *Appl Environ Microbiol* 2007;**73**:4089–93.

52. Senesi S, Celandroni F, Scher S, Wong ACL, Ghelardi E. Swarming motility in *Bacillus cereus* and characterization of a *fliY* mutant impaired in swarm cell differentiation. *Microbiology* 2002;**148**:1785–94.

53. Henrichsen J. Bacterial surface translocation: a survey and classification. *Bacteriol Rev* 1972;**36**:478–503.

54. Allison C, Coleman N, Jones PL, Hughes C. Ability of *Proteus mirabilis* to invade human uroepithelial cells is coupled to motility and swarming differentiation. *Infect Immun* 1992;**60**:4740–6.

55. Moyer AL, Ramadan RT, Novosad BD, Astley R, Collagen MC. *Bacillus cereus*-induced permeability of the blood ocular barrier during experimental endophthalmitis. *Invest Ophthalmol Vis Sci* 2009;**50**:3783–93.

56. Chen M, Hou PK, Tai TU, Lin BJ. Blood-ocular barriers. *Tzuchi Med J* 2008;**20**:25–34.

57. Kopel AC, Carvounis PE, Holz ER. *Bacillus cereus* endophthalmitis following invitreous bevacizumab injection. *Ophthalmic Surg Lasers Imaging* 2008;**39**:153–4.

58. Lebessi E, Dellagrammaticas HD, Antonaki G, Foustoukou M, Iacovidou N. *Bacillus cereus* meningitis in a term neonate. *J Matern Fetal Neonatal Med* 2009;**22**:458–61.

59. Turnbull PCB, Jorgensen K, Kramer JM, Gilbert RJ, Perry JM. Severe clinical conditions associated with *Bacillus cereus* and the apparent involvements of exotoxins. *J Clin Pathol* 1979;**32**:289–93.

60. Marley EF, Saini NK, Venkatraman C, Orenstein JM. Fatal *Bacillus cereus* meningoencephalitis in an adult with myelogenous leukemia. *South Med J* 1995;**88**:969–72.

61. Akiyama N, Mitani K, Tanaka Y, Hanazono Y, Motoi N, Zarkovic M, et al. Fulminant septicemic syndrome of *Bacillus cereus* in a leukemic patient. *Intern Med* 1997;**36**:221–6.

62. Funada H, Votani C, Machi T, Matsuda T, Nonomura A. *Bacillus cereus* bacteremia in an adult with acute leukemia. *Jpn J Clin Oncol* 1988;**18**:69–74.

63. Kawatani E, Kishikawa Y, Sankoda C, Kuwahara N, Mori D, Osoegawa K, et al. *Bacillus cereus* sepsis and subarachnoid hemorrhage following consolidation chemotherapy for acute myelogenous leukemia. *Rinsho Ketsueki* 2009;**50**:300–3. [in Japanese].

64. Ihde DC, Armstrong D. Clinical spectrum of infection due to *Bacillus* species. *Am J Med* 1973;**55**:839–45.

65. Saki C, Iuchi T, Ishii A, Kumagai K, Takagi T. *Bacillus cereus* brain abssesses occurring in severely neutropenic patients: successful treatment with antimicrobial agents, granulocyte colony-stimulating factor, and surgical drainage. *Intern Med* 2001;**40**:654–7.

66. Tanabe T, Kodama Y, Nisaikawa T, Okamoto Y, Kawano Y. Critical illness polyneuropathy after *Bacillus cereus* sepsis in acute lymphoblastic leukemia. *Intern Med* 2005;**48**:1175–7.

67. Gaur AH, Patrick CC, McCullers JA, Flynn PA, Pearsons TA, Razzouk BI, Thompson SJ, Shenep JL. *Bacillus cereus* bacteremia and meningitis in immunocompromised children. *Clin Infect Dis* 2001;**32**:1456–62.

68. Manickam N, Knorr A, Muldrew KL. Neonatal meningoencephalitis caused by *Bacillus cereus*. *Pediatr Infect Dis J* 2008;**27**:843–5.

69. Arnaout MK, Tamburro RT, Bodner SM, Sandlund JT, Rivera GK, Pui CH, et al. *Bacillus cereus* fulminant sepsis and hemolysis in two patients with acute leukemia. *J Pediatr Hematol Oncol* 1999;**21**:431–5.

70. Jenson HB, Levy SR, Duncan C, McIntosh S. Treatment of multiple brain abscesses caused by *Bacillus cereus*. *Pediatr Infect Dis J* 1989;**8**:795–8.

71. Yoshida H, Moriyama Y, Tatekawa T, Tominaga T, Teshima H, Hiraoka H, et al. Two cases of acute myelogenous leukemia with *Bacillus cereus* bacteremia resulting in fatal intracranial hemorrhage. *Rinsho Ketsueki* 1993;**34**:1568–72. [in Japanese].

72. Ghosh AC. Prevalence of *Bacillus cereus* in the faeces of healthy adults. *J Hyg (Lond)* 1978;**80**:233–6.

73. Girsch M, Ries M, Zenker M, Carbon R, Rauch A, Hofbeck M. Intestinal perforation in a premature infant caused by *Bacillus cereus*. *Infection* 2003;**31**:192–3.

74. Le Scanff J, Mohammedi JI, Thiebaut A, Martin O, Argaud L, Robert D. Necrotizing gastroenteritis due to *Bacillus cereus* in an immunocompromised patient. *Infection* 2006;**34**:98–9.

75. Lund T, Granum PE. Comparison of biological effect of two different enterotoxin complexes isolated from three different strains of *Bacillus cereus*. *Microbiology* 1997;**143**:3329–36.

76. Colpin GGD, Guiot HFL, Simonis RFA, Zwann EE. *Bacillus cereus* meningitis in a patient under gnotobiotic care. *Lancet* 1981;**ii**:694–5.

77. Guiot HFL, de Planque MM, Richel DJ, Van't Wout JW. *Bacillus cereus*: a snake in the grass for granulocytopenic patients. *J Infect Dis* 1986;**153**:1186.

78. Darbar A, Harris IA, Gosbell IB. Necrotizing infection due to *Bacillus cereus* mimicking gas gangrene following penetrating trauma. *J Orthop Trauma* 2005;**19**:353–5.

79. Dubouix A, Bonnet E, Bensafi MH, Archambaud M, Chaminade B, Echabanon G, et al. *Bacillus cereus* infections in traumatology orthopedics department: retrospective investigation and improvement of health practices. *J Infect* 2005;**50**:22–30.

80. Flores-Diaz M, Alape-Giron A. Role of *Clostridium perfringens* phospholipase C in the pathogenesis of gas gangrene. *Toxicon* 2008;**42**:979–86.

81. Gröschel M, Burges A, Bodey GP. Gas gangrene-like infection with *Bacillus cereus* in a lymphoma patient. *Cancer* 1976;**37**:988–91.

82. Bessman AN, Wagner W. Nonclostridial gas gangrene. Report of 48 cases and review of the literature. *JAMA* 1975;**42**:958–63.

83. Ribeiro NFF, Heath CH, Kierath J, Rea S, Duncan-Smith M, Wood FM. Burn wounds infected by contaminated water: case reports, review of the literature, and recommendations for treatment. *Burns* 2010;**36**:9–22.

84. Äkesson A, Hedstrőm SA, Ripa TE. *Bacillus cereus*: a significant pathogen in postoperative and post-traumatic wounds in orthopaedic wards. *Scand J Infect Dis* 1991;**23**:22–30.

85. Brett MM, Hood J, Brazier JS, Duerden BI, Hahné JM. Soft tissue infections by spore-forming bacteria in injecting drug users in the United Kingdom. *Epidemiol Infect* 2005;**133**:575–82.

86. Krause A, Freeman A, Sisson PA, Murphy OM. Infection with *Bacillus cereus* after close-range gunshot injuries. *J Trauma Injury Infect Crit Care* 1996;**41**:546–7.

87. Hernandez E, Dubrous P, Deloynes B, Cavallo JD. Infections and superinfections due to *Bacillus cereus* following severe war injuries of French soldiers in former Yugoslavia and the Gulf War, abstr. K-110. Abstr. In: *36th Intersci. conf. antimicrob. agents chemother.* Washington (DC): American Society for Microbiology; 1996.

88. Rutala WA, Saviteer SM, Thomann CA, Wilson MB. Plaster-associated *Bacillus cereus* wound infection. A case report. *Orthopedics* 1986;**9**:575–7.

89. Henrickson KJ. A second species of *Bacillus* causing primary cutaneous disease. *Int J Dermatol* 1990;**29**:19–20.

90. Khavari PA, Bolognia JL, Eisen R, Edberg SC, Grimshaw SC, Shapiro PE. Periodic acid-Schiff-positive organisms in primary cutaneous *Bacillus cereus* infections. *Arch Dermatol* 1991;**127**:543–6.

91. Henrickson KJ, Flynn PM, Shenep JL, Pui CH. Primary cutaneous *Bacillus cereus* infections in neutropenic children. *Lancet* 1989;**i**:601–3.

92. Heyworth B, Ropp ME, Voos UG, Meinel HI, Darlow HM. Anthrax in the Gambia: an epidemiological study. *Br Med J* 1975;**iv**:79–82.

93. Beecher DJ, Schoeni JL, Wong ACL. Enterotoxic activity of hemolysin BL from *Bacillus cereus*. *Infect Immun* 1995;**63**:4423–8.

94. Steen MK, Bruno-Murtha LA, Chaux G, Lazar H, Bernard S, Sulis C. *Bacillus cereus* endocarditis: report of a case and review. *Clin Infect Dis* 1992;**14**:945–6.

95. Block CS, Levy ML, Fritz VU. *Bacillus cereus* endocarditis. *S Afr Med J* 1978;**53**:556–7.

96. Castedo E, Castro A, Martin P, Roda J, Montero CG. *Bacillus cereus* prosthetic valve endocarditis. *Ann Thorac Surg* 1999;**68**:2351–2.

97. Oster HA, Kong TQ. *Bacillus cereus* endocarditis involving a prosthetic valve. *South Med J* 1982;**75**:508–9.

98. Sliman R, Rehm S, Shlaes D. Serious infections caused by *Bacillus* species. *Medicine (Baltimore)* 1987;**66**:218–23.

99. Abusin S, Bhimaraj A, Khadra S. *Bacillus cereus* endocarditis in a permanent pacemaker: a case report. *Cases J* 2008;**1**:95.

100. Cone LA, Dreisbach L, Potts BE, Comess BE, Burleigh WA. Fatal *Bacillus cereus* endocarditis masquerading as an anthrax-like infection in a patient with acute lymphoblastic leukemia: case report. *J Heart Valve Dis* 2005;**14**:37–9.

101. Tomomasa T, Itoh K, Matsui A, Kobayashi T, Suzuki N, Matsuyama S, et al. An infant with ulcerative colitis complicated by endocarditis and cerebral infarction. *J Pediatr Gasterenterol Nutr* 1993;**17**:323–5.

102. Schricker ME, Thompson GH, Schreiber JR. Osteomyelitis due to *Bacillus cereus* in an adolescent: case report and review. *Clin Infect Dis* 1994;**18**:863–7.

103. Farrar WE. Serious infections due to "non-pathogenic" organisms of the genus *Bacillus*. *Am J Med* 1963;**34**:134–41.

104. Sato K, Ichiyama S, Ohmura M, Takashi M, Agata N, Ohta M, et al. A case of urinary tract infection caused by *Bacillus cereus*. *J Infect* 1998;**36**:247–8.

# CHAPTER 9

# *Bacillus cereus* Disease in Animals

Vincenzo Savini

Clinical Microbiology and Virology, Laboratory of Bacteriology and Mycology, Civic Hospital of Pescara, Pescara, Italy

## SUMMARY

Although cattle mastitis is mostly due to staphylococci, streptococci, and *Escherichia coli*, *Bacillus cereus* may behave as an agent of mammary gland infection in cows and goats. Also, *Bacillus* species outside the *B. cereus* group seem to be potential pathogens for the mammalian udder, although many cases are merely reported as related to *Bacillus* spp., unfortunately. Finally, *Bacillus anthracis* still remains the major pathogen within the genus, not only in humans, but also in animals, in which it frequently leads the affected host to death.

## BOVINES

It is known that *B. cereus* is frequently isolated from dairy foods. Particularly, and commonly, psychrotolerant strains of *Bacillus* species in general are found in pasteurized milk and milk products.[1–4] *Bacillus cereus*, additionally, may be responsible for off-flavors in milk even at low counts, as well as for the defect known as "bitty cream."[5] Such data clearly emphasize the role of animals as a potential source of *B. cereus*; in fact, and in particular, the organism may cause mammalian mastitis, both in bovines and in goats, thus contaminating milk and its derivatives at all stages of processing.[6–9]

Although bovine mastitis is generally due to organisms such as *Staphylococcus aureus*, streptococci, and *E. coli*, which represent the major udder pathogens, uncommonly encountered bacteria may cause occasional cases of mammary gland infection. Among these unusual organisms, *B. cereus* can be responsible for marked tissue damage and grossly abnormal secretion from the udder.[6–9]

In general, the clinical picture in cows includes fever, bloodstained milk, uniformly bloody udder secretions, and, sometimes, the presence of necrotic mammary gland tissue in the milk. Again, the involved udder is

*The Diverse Faces of Bacillus cereus*
ISBN 978-0-12-801474-5
http://dx.doi.org/10.1016/B978-0-12-801474-5.00009-8

enlarged, hard, and painful to the touch and it shows skin discoloration. Ill cows are almost completely inappetant, are lethargic, and may exhibit signs of shock. Udder necrosis is usually observed over a short period of time, and a substantial reduction in the milk yield from the unaffected quarters may be documented in lactating bovines. Mild cases may be characterized by the presence of clots in the milk and moderate flogosis of the affected quarter and usually respond promptly to intramammary antibiotics; gangrenous infections instead may lead to necrosis and sloughing of the involved quarter, with extensive bluish-purple discoloration of the skin overlying the quarter itself.[6] In general, over time, affected quarters can slough, and one can observe a return to milk production in the unaffected quarters. Nevertheless, milk production in the current and future lactations is generally reduced.[6–9]

*Bacillus cereus* is known to be ubiquitous in the farm environment, and the spore amount in the soil usually rises in winter.[10] Consequently, although the bacterium is not labeled as a primary mastitis pathogen, it may infect the mammary gland through accidental introduction into the udder, given its ubiquity, and bovine diseases usually follow contaminated teat surgery.[6–9,11] Also, in cattle, mastitis has been reported to occur after the administration of contaminated intramammary antibiotics and to be observed at the time of the next calving. Otherwise, cases of *B. cereus* mastitis in bovines have been related to contaminated concentrate feeds, notably brewers' grains, as well as contaminated syringes, teat tubes, and teat dilators. Contamination of chaff and feeds, also, might be involved. Fatal outcomes have been reported in bovines, although most animals usually, and fortunately, survive.[6,10–12]

Notably, systemic alterations due to gangrenous *B. cereus* mastitis include disseminated intravascular coagulation, hemolysis, and hemoglobinuria, with related hemoglobinuric nephrosis.[12]

Pathological alterations observed in *B. cereus* mastitis are related to the tripartite exotoxin, exerting cytotoxic and dermonecrotic effects, as well as causing vascular permeabilization, while the mentioned systemic signs rely on a variety of toxins, including phospholipases and hemolysins.[6]

## GOATS

*Bacillus cereus* mastitis has been reported in goats, as well, which show a clinical syndrome that is similar to that observed in cows. Gangrenous infection due to this organism has been observed to cause goat fever and inappetence, lethargy, and dehydration. The affected udder is firm and warm, and hemorrhagic milk secretion is usually observed to come out from the pathologic gland.[12] Additional signs are tachycardia and tachypnea, bilateral scleral injection, leukopenia with band neutrophils, mild azotemia, hypocalcemia, and decreased bicarbonate concentration.[12] A clear demarcation may be visible between the half normal skin color and the half darker reddish-brown skin discoloration in the affected area. Antibiotic and antiflogistic treatment usually leads to slow resolution some months from the disease onset, with sloughing of the affected gland, wound healing, and growth of normal haired skin covering the affected zone.[12]

In general, in lactations following mastitis resolution, milk production may be above the mean; otherwise goats are voluntarily withdrawn from lactation. Information from the literature, however, is confused sometimes, as identification at the species level has not been performed in all described cases, for which the mastitis episodes were in fact simply labeled as *Bacillus*-related. Goats, however, usually seem to have successful kiddings after the infection, and prognosis for survival is good when proper treatment is started promptly.[12]

## VETERINARY ANTHRAX

*Bacillus cereus* is ubiquitous in the environment and behaves therefore as a common milk contaminant, whereas mastitis caused by this organism is considered to be rare, and specific predisposing factors are supposed to exist behind this syndrome.[12–16] Being that *B. cereus* is an environmental contaminant, as said, the diagnosis of *B. cereus* mastitis, of course, relies on the isolation of the organism from a correctly collected milk sample from an animal with signs of endotoxemia and gangrenous changes in an udder.[12]

Notably, aside from mammary gland infection, *B. cereus* has been known to cause abortion in cattle, as confirmed experimentally by intravenous inoculation in heifers and sheep, and, finally, it may be of interest to note that *B. cereus* serotypes from bovine mastitis seem to be different from those collected from humans.[12]

Unfortunately, as mentioned previously, a number of clinical cases of mastitis have been attributed to *Bacillus* species, without any identification being performed at the species level, and the number of subclinical mastitis cases due to *Bacillus* is unknown.[17] In particular, heat-stable toxin-producing *Bacillus licheniformis* and *Bacillus pumilus* have been isolated from the milk of bovines with mastitis. Spores produced by these species are known to be heat stable and to resist hydrogen peroxide; thus they may survive cleaning protocols and dairy processes such as spray evaporation.[17] Milk from subclinical *B. licheniformis* and *B. pumilus* carriers may therefore introduce a risk to the safety of milk-powder products.[17–52]

Notably, outside the genus *Bacillus*, the name *Streptococcus agalactiae*, meaning "no milk," was given to that organism as it was originally described as causing bovine mastitis.[53]

Animal mastitis is in fact mostly related to infection by *Staphylococcus* species, as well as *E. coli* and streptococci, while to a lesser extent this syndrome may be traced back to a *Bacillus* infection.[18–58] Nevertheless, *B. cereus*, and *Bacillus* species even outside the *B. cereus* group, should be included among the potential etiologic agents of mammary gland infection in bovines and goats.

Within the genus *Bacillus*, however, *B. anthracis* is the major agent of disease, in general, like in humans. It is a free-living bacterium usually present as spores in alkaline soils. Spores are the resistant and avirulent forms of the pathogen and can live in the soil for numerous decades and still be viable when they come into contact with a susceptible host. The bacterium produces edema and lethal toxins based on a combination of its three virulence factors, named protective antigen, lethal factor, and edema factor. Virulence is further increased by the antiphagocytic capsular antigen. The toxins are responsible for the primary clinical signs of necrosis, edema, and hemorrhage that are typically observed in infected hosts.[59–66]

Anthrax is characterized by a high fatality rate in herbivores and other susceptible hosts and mostly leads to animal death. Susceptibility and high lethality have been documented in sheep, goats, cattle, horses, donkeys, and swine, together with many other warm-blooded domestic animals. Wildlife with high rates of anthrax include, moreover, antelopes, bisons, gazelles, and impalas, but also elephants and hippopotami. Finally, wild carnivores can become infected as well, via consumption of anthrax-infected dead animals. Birds have instead a natural resistance to anthrax and are not able to carry *B. anthracis* to uninfected locations. Notably, outbreaks have been observed in animals that had ingested feedstuffs containing meat and bonemeal-based concentrates whose origins were carcasses contaminated with anthrax spores.[67–72]

Exhaustion of nutrients, oxygen exposure, and death of host tissues may make *B. anthracis* sporulate. Spores return to the soil with the burial, decomposition, or rupture of the infected carcass and can be subsequently picked up by other animals by feeding. The avirulent phase can last several years; then, when soil conditions become newly favorable, the spores migrate through capillary action of soil water to the surface. Anthrax spores may adhere to plants as the vegetation resurfaces during evaporation. In enzootic zones, when animals graze close to one another on fresh shoots of grass after high rainfall, outbreaks may occur.[68–73]

Although the major mechanism of anthrax transmission is ingestion of infective bacteria, biting flies may transmit the disease between animals. Nonbiting blowflies instead contaminate vegetation by virtue of deposited vomit droplets after feeding on infected carcasses. In their turn, animals feeding on such vegetation consequently have a higher chance of acquiring the infection. Another vehicle is represented by blood-feeding insects. Of concern, agricultural areas where veterinary public health facilities are inadequate show the highest occurrence of animal anthrax. The incidence, however, is unknown in several countries, but it is likely that the organism is present in most regions.[68–78]

Antibiotic administration is recommended in the literature as a treatment against anthrax both in humans and in animals. In the latter, penicillin and tetracycline are usually effective, although there seems to be evidence that an antagonistic effect exists between the vaccine and antibiotic administration during the immunogenic period.[79,80] Again, enhanced resistance to antibiotics is a further matter of concern that needs to be addressed in the ambit of antimicrobial therapy in animal anthrax. It has also been conclusively documented that antibiotic treatment during an infection has prevented the establishment of an immune response in some animals and, despite the antibiotic's effectiveness, there exists the risk of recurring disease because of delayed spore germination.[79,80]

In conclusion, in addition to freely living as ubiquitous microorganisms or behaving as human pathogens, *Bacillus* species may play a role as agents of infectious diseases in the veterinary setting. Changes are required in bacterial characterization methodologies to achieve *Bacillus* identification at the species level and to make such methods more and more available in the veterinary and medical diagnostic microbiology laboratories; it is hoped that this will gradually lead to an increasing consciousness of the pathogenic potential in animals of each organism of the *B. cereus* group, as well as *Bacillus* species outside of it.

## REFERENCES

1. Christiansson A, Naidu AS, Nilsson I, Wadstrom T, Pettersson H. Toxin production by *Bacillus cereus* dairy isolates in milk at low temperatures. *Appl Environ Microbiol* 1989;**55**: 2595–600.
2. Meer RR, Baker J, Bodyfelt FW, Griffiths MW. Psychrotrophic *Bacillus* spp. in fluid milk products - a review. *J Food Prot* 1991;**54**:969–79.
3. Vaisanen OM, Mwaisumo NJ, Salkinoja-Salonen MS. Differentiation of dairy strains of the *Bacillus cereus* group by phage typing, minimum growth temperature, and fatty acid analysis. *J Appl Bacteriol* 1991;**70**:315–24.

4. van Netten P, van De Moosdijk A, van Hoensel P, Mossel DA, Perales I. Psychrotrophic strains of *Bacillus cereus* producing enterotoxin. *J Appl Bacteriol* 1990;**69**:73–9.
5. Stenfors LP, Mayr R, Scherer S, Granum PE. Pathogenic potential of fifty *Bacillus weihenstephanensis* strains. *FEMS Microbiol Lett* 2002;**215**:47–51.
6. Parkinson TJ, Merrall M, Fenwick SG. A case of bovine mastitis caused by *Bacillus cereus*. *N Z Vet J* 1999;**47**:151–2.
7. Crielly EM, Logan NA, Anderton A. Studies on the *Bacillus* flora of milk and milk products. *J Appl Bacteriol* 1994;**77**:256–63.
8. Jones TO, Turnbull PC. Bovine mastitis caused by *Bacillus cereus*. *Vet Rec* 1981;**108**: 271–4.
9. Perrin D, Greenfield J, Ward GE. Acute *Bacillus cereus* mastitis in dairy cattle associated with use of a contaminated antibiotic. *Can Vet J* 1976;**17**:244–7.
10. Davies RH, Wray C. Seasonal variations in the isolation of *Salmonella typhimurium*, *Salmonella enteritidis*, *Bacillus cereus* and *Clostridium perfringens* from environmental samples. *Zentralbl Veterinarmed B* 1996;**43**:119–27.
11. Scheifer B, Macdonald KR, Klavano GG, van Dreumel AA. Pathology of *Bacillus cereus* mastitis in dairy cows. *Can Vet J* 1976;**17**:239–43.
12. Mavangira V, Angelos JA, Samitz EM, Rowe JD, Byrne BA. Gangrenous mastitis caused by *Bacillus* species in six goats. *J Am Vet Med Assoc* 2013;**242**:836–43.
13. Beecher DJ, Schoeni JL, Wong AC. Enterotoxic activity of hemolysin BL from *Bacillus cereus*. *Infect Immun* 1995;**63**:4423–8.
14. Graham C. *Bacillus species* and non-spore-forming anaerobes in New Zealand livestock. *Surveillance* 1998;**24**:19.
15. Johnson KG. Bovine mastitis caused by *Bacillus cereus*. *Vet Rec* 1981;**108**:404–5.
16. Logan NA. *Bacillus* species of medical and veterinary importance. *J Med Microbiol* 1988;**25**:157–65.
17. Nieminen T, Rintaluoma N, Andersson M, Taimisto AM, Ali-Vehmas T, Seppälä A, et al. Toxinogenic *Bacillus pumilus* and *Bacillus licheniformis* from mastitic milk. *Vet Microbiol* 2007;**124**:329–39.
18. Weese JS, van Duijkeren E. Methicillin-resistant *Staphylococcus aureus* and *Staphylococcus pseudintermedius* in veterinary medicine. *Vet Microbiol* 2010;**140**:418–29.
19. Peton V, Le Loir Y. *Staphylococcus aureus* in veterinary medicine. *Infect Genet Evol* 2014;**21**:602–15.
20. Bergonier D, de Crémoux R, Rupp R, Lagriffoul G, Berthelot X. Mastitis of dairy small ruminants. *Vet Res* 2003;**34**:689–716.
21. Le Loir Y, Baron F, Gautier M. *Staphylococcus aureus* and food poisoning. *Genet Mol Res* 2003;**2**:63–76.
22. Le Maréchal C, Seyffert N, Jardin J, Hernandez D, Jan G, Rault L, et al. Molecular basis of virulence in *Staphylococcus aureus* mastitis. *PLoS One* 2011;**6**:e27354.
23. Guinane CM, Sturdevant DE, Herron-Olson L, Otto M, Smyth DS, Villaruz AE, et al. Pathogenomic analysis of the common bovine *Staphylococcus aureus* clone (ET3): emergence of a virulent subtype with potential risk to public health. *J Infect Dis* 2008;**197**:205–13.
24. Zhao X, Lacasse P. Mammary tissue damage during bovine mastitis: causes and control. *J Anim Sci* 2008;**86**:57–65.
25. Pereira UP, Oliveira DG, Mesquita LR, Costa GM, Pereira LJ. Efficacy of *Staphylococcus aureus* vaccines for bovine mastitis: a systematic review. *Vet Microbiol* 2011;**148**:117–24.
26. Klostermann K, Crispie F, Flynn J, Ross RP, Hill C, Meaney W. Intramammary infusion of a live culture of *Lactococcus lactis* for treatment of bovine mastitis: comparison with antibiotic treatment in field trials. *Dairy Res* 2008;**75**:365–73.
27. Bouchard DS, Rault L, Berkova N, Le Loir Y, Even S. Inhibition of *Staphylococcus aureus* invasion into bovine mammary epithelial cells by contact with live *Lactobacillus casei*. *Appl Environ Microbiol* 2013;**79**:877–85.

28. Rupp R, Bergonier D, Dion S, Hygonenq MC, Aurel MR, Robert-Granié C, et al. Response to somatic cell count-based selection for mastitis resistance in a divergent selection experiment in sheep. *J Dairy Sci* 2009;**92**:1203–19.

29. Wall RJ, Powell AM, Paape MJ, Kerr DE, Bannerman DD, Pursel VG, et al. Genetically enhanced cows resist intramammary *Staphylococcus aureus* infection. *Nat Biotechnol* 2005;**23**:445–51.

30. Hajek V. *Staphylococcus intermedius*, a new species isolated from animals. *Int J Syst Bacteriol* 1976;**26**:401–8.

31. Devriese LA, Vancanneyt M, Baele M, Vaneechoutte M, De Graef E, Snauwaert C, et al. *Staphylococcus pseudintermedius* sp. nov., a coagulase-positive species from animals. *Int J Syst Evol Microbiol* 2005;**55**:1569–73.

32. Sasaki T, Kikuchi K, Tanaka Y, Takahashi N, Kamata S, Hiramatsu K. Reclassification of phenotypically identified *Staphylococcus intermedius* strains. *J Clin Microbiol* 2007;**45**:2770–8.

33. Savini V, Passeri C, Mancini G, Iuliani O, Marrollo R, Argentieri AV, et al. Coagulase-positive staphylococci: my pet's two faces. *Res Microbiol* 2013;**164**:371–4.

34. Savini V, Barbarini D, Polakowska K, Gherardi G, Bialecka A, Kasprowicz A, et al. Methicillin-resistant *Staphylococcus pseudintermedius* infection in a bone marrow transplant recipient. *J Clin Microbiol* 2013;**51**:1636–8.

35. Bannoehr J, Ben Zakour NL, Waller AS, Guardabassi L, Thoday KL, van den Broek AH, et al. Population genetic structure of the *Staphylococcus intermedius* group: insights into agr diversification and the emergence of methicillin-resistant strains. *J Bacteriol* 2007;**189**: 8685–92.

36. Devriese LA, Hermans K, Baele M, Haesebrouck F. *Staphylococcus pseudintermedius* versus *Staphylococcus intermedius*. *Vet Microbiol* 2009;**133**:206–7.

37. Ruscher C, Lübke-Becker A, Wleklinski CG, Soba A, Wieler LH, Walther B. Prevalence of Methicillin-resistant *Staphylococcus pseudintermedius* isolated from clinical samples of companion animals and equidaes. *Vet Microbiol* 2009;**136**:197–201.

38. van Duijkeren E, Kamphuis M, van der Mije IC, Laarhoven LM, Duim B, Wagenaar JA, et al. Transmission of methicillin-resistant *Staphylococcus pseudintermedius* between infected dogs and cats and contact pets, humans and the environment in households and veterinary clinics. *Vet Microbiol* 2011;**150**:338–43.

39. Talan DA, Staatz D, Staatz A, Overturf GD. Frequency of *Staphylococcus intermedius* as human nasopharyngeal flora. *J Clin Microbiol* 1989;**27**:2393.

40. Mahoudeau I, Delabranche X, Prevost G, Monteil H, Piemont Y. Frequency of isolation of *Staphylococcus intermedius* from humans. *J Clin Microbiol* 1997;**35**:2153–4.

41. Talan DA, Goldstein EJ, Staatz D, Overturf GD. *Staphylococcus intermedius*: clinical presentation of a new human dog bite pathogen. *Ann Emerg Med* 1989;**18**:410–3.

42. Guardabassi L, Loeber ME, Jacobson A. Transmission of multiple antimicrobial-resistant *Staphylococcus intermedius* between dogs affected by deep pyoderma and their owners. *Vet Microbiol* 2004;**98**:23–7.

43. Lee J. *Staphylococcus intermedius* isolated from dog-bite wounds. *J Infect* 1994;**29**:105.

44. Tanner MA, Everett CL, Youvan DC. Molecular phylogenetic evidence for noninvasive zoonotic transmission of *Staphylococcus intermedius* from a canine pet to a human. *J Clin Microbiol* 2000;**38**:1628–31.

45. Sasaki T, Kikuchi K, Tanaka Y, Takahashi N, Kamata S, Hiramatsu K. Methicillin-resistant *Staphylococcus pseudintermedius* in a veterinary teaching hospital. *J Clin Microbiol* 2007;**45**:1118–25.

46. van Duijkeren E, Houwers DJ, Schoormans A, Broekhuizen-Stins MJ, Ikawaty R, Fluit AC, et al. Transmission of methicillin-resistant *Staphylococcus intermedius* between humans and animals. *Vet Microbiol* 2008;**128**:213–5.

47. Simojoki H, Orro T, Taponen S, Pyörälä S. Host response in bovine mastitis experimentally induced with *Staphylococcus chromogenes*. *Vet Microbiol* 2009;**134**:95–9.

48. Ruegg PL. The quest for the perfect test: phenotypic versus genotypic identification of coagulase-negative staphylococci associated with bovine mastitis. *Vet Microbiol* 2009;**134**: 15–9.

49. Åvall-Jääskeläinen S, Koort J, Simojoki H, Taponen S. Bovine-associated CNS species resist phagocytosis differently. *BMC Vet Res* 2013;**9**:227.

50. Piette A, Verschraegen G. Role of coagulase-negative staphylococci in human disease. *Vet Microbiol* 2009;**134**:45–54.

51. Gandolfi-Decristophoris P, Regula G, Petrini O, Zinsstag J, Schelling E. Prevalence and risk factors for carriage of multi-drug resistant staphylococci in healthy cats and dogs. *J Vet Sci* 2013;**14**:449–56.

52. Weese JS, Dick H, Willey BM, McGeer A, Kreiswirth BN, Innis B, et al. Suspected transmission of methicillin-resistant *Staphylococcus aureus* between domestic pets and humans in veterinary clinics and in the household. *Vet Microbiol* 2006;**115**:148–55.

53. Savini V, Gherardi G, Marrollo R, Franco A, Pimentel De Araujo F, Dottarelli S, et al. Could β-hemolytic, group B *Enterococcus faecalis* be mistaken for *Streptococcus agalactiae*? *Diagn Microbiol Infect Dis* 2014;**82**(1):32–3. [Epub ahead of print].

54. Savini V, Marrollo R, Coclite E, Fusilli P, D'Incecco C, Fazii P. Successful off-label use of the Cepheid Xpert GBS in a late-onset neonatal meningitis by *Streptococcus agalactiae*. *Int J Clin Exp Pathol* 2014;**7**(8):5192–5.

55. Savini V, Franco A, Gherardi G, Marrollo R, Argentieri AV, Pimentel de Araujo F, et al. Beta-hemolytic, multi-lancefield antigen-agglutinating *Enterococcus durans* from a pregnant woman, mimicking *Streptococcus agalactiae*. *J Clin Microbiol* 2014;**52**(6):2181–2.

56. Savini V, Marrollo R, D'Antonio M, D'Amario C, Fazii P, D'Antonio D. *Streptococcus agalactiae* vaginitis: nonhemolytic variant on the Liofilchem® Chromatic StreptoB. *Int J Clin Exp Pathol* 2013;**6**(8):1693–5.

57. Savini V, Santarelli A, Polilli E, D'Antonio M, Astolfi D, Marrollo R, et al. Looking for nonhemolytic group B streptococci. *Vet Microbiol* 2013;**163**(1–2):204–5.

58. Savini V. *A visit from the stork: infants and GBS*. LAP Lambert Academic Publishing; 2015.

59. Murray PR, Pfaller MA, Rosenthal KS. *Medical microbiology*. 5th ed. Philadelphia: Mosby; 2005.

60. Ryu C, Lee K, Seong WK, Oh HB. Sensitive and rapid quantitative detection of anthrax spores isolated from soil samples by real-time PCR. *Microbiol Immunol* 2003;**47**:693–9.

61. Dragon DC, Rennie RP. The ecology of anthrax spores: tough but not invincible. *Can Vet J* 1995;**36**:295–301.

62. Ryan KJ, Ray CG, editors. *Sherris medical microbiology*. 4th ed. New York: McGraw Hill; 2004.

63. World Health Organization. *Guidelines for the surveillance and control of anthrax in humans and animals*. 1998. Available from: http://www.who.int/csr/resources/publications/anthrax/WHO_EMC_ZDI_98_6/en. [cited 14.02.07].

64. Ramisse V, Patra G, Garrigue H, Guesdon JL, Mock M. Identification and characterization of *Bacillus anthracis* by multiplex PCR analysis of sequences on plasmids pX01 and pX02 and chromosomal DNA. *FEMS Microbiol Lett* 1996;**145**:9–16.

65. Hanna PC, Ireland JA. Understanding *Bacillus anthracis* pathogenesis. *Trends Microbiol* 1999;**7**:180–2.

66. The virulence of cultivated anthrax virus. *Science* 1884;**4**:276–7.

67. Bales ME, Dannenberg AL, Brachman PS, Kaufmann AF, Klatsky PC, Ashford DA. Epidemiologic response to anthrax outbreaks: field investigations, 1950–2001. *Emerg Infect Dis* 2002;**8**:1163–74.

68. U.S. Department of Agriculture: animal and plant health inspection service, veterinary services, centers for epidemiology and animal health, center for emerging issues (US) epizootiology and ecology of anthrax. Available from: http://www.aphis.usda.gov/vs/ceah/cei/taf/emerginganimalhealthissues_files/anthrax.text.pdf; [cited 14.02.07].

69. Lindeque PM, Turnbull PC. Ecology and epidemiology of anthrax in Etosha National Park, Namibia. *Onderstepoort J Vet Res* 1994;**61**:71–83.
70. Food and Agriculture Organization of the United Nations. Agriculture and Consumer Protection Department. Anthrax in animals. Available from: http://www.fao.org/ag/magazine/0112sp.htm; [cited 14.02.07].
71. Human anthrax associated with an epizootic among livestock—North Dakota, 2000. *Morb Mortal Wkly Rep* 2001;**50**:677–80.
72. Human ingestion of *Bacillus anthracis*-contaminated meat—Minnesota, August 2000. *JAMA* 2000;**284**:1644–6.
73. Titball RW, Turnbull PC, Hutson RA. The monitoring and detection of *Bacillus anthracis* in the environment. *Soc Appl Bacteriol Symp Ser* 1991;**20**:S9–18.
74. Turnbull PC, Lindeque PM, Le Roux J, Bennett AM, Parks SR. Airborne movement of anthrax spores from carcass sites in the Etosha National Park, Namibia. *J Appl Microbiol* 1998;**84**:667–76.
75. Turell MJ, Knudson GB. Mechanical transmission of *Bacillus anthracis* by stable flies (*Stomoxys calcitrans*) and mosquitoes (*Aedes aegypti* and *Aedes taeniorhynchus*). *Infect Immun* 1987;**55**:1859–61.
76. Friedlander AM. Microbiology: tackling anthrax. *Nature* 2001;**414**:160–1.
77. Dragon DC, Bader DE, Mitchell J, Wollen N. Natural dissemination of *Bacillus anthracis* spores in northern Canada. *Appl Environ Microbiol* 2005;**71**:1610–5.
78. Madigan M, Martinko J. *Brock biology of microorganisms*. 11th ed. Carbondale (IL): Prentice Hall; 2005.
79. Inglesby TV, O'Toole T, Henderson DA, Bartlett JG, Ascher MS, Eitzen E, et al. Anthrax as a biological weapon, 2002: updated recommendations for management. *JAMA* 2002;**287**:2236–52.
80. Friedlander AM, Welkos SL, Pitt ML, Ezzell JW, Worsham PH, Rose KJ, et al. Postexposure prophylaxis against experimental inhalation anthrax. *J Infect Dis* 1993;**167**:1239–43.

# CHAPTER 10

# *Bacillus cereus* Biocontrol Properties

## Vincenzo Savini
Clinical Microbiology and Virology, Laboratory of Bacteriology and Mycology,
Civic Hospital of Pescara, Pescara, Italy

## SUMMARY

Zwittermicin A represents a new antibiotic class and is produced by *Bacillus cereus*; it shows diverse biological activities, such as suppression of microbial diseases in plants along with potentiation of the insecticidal effectiveness of *Bacillus thuringiensis*.

## BIOCONTROL IN *BACILLUS CEREUS*

It has been known that the virtuosity of bacteria as synthetic chemists is unparalleled in nature. In fact, they may produce diverse compounds exerting a broad spectrum of activities; these have been used by humans for several purposes over the years. The need for new antimicrobial molecules has increased in recent decades owing to the emergence of new pathogenic microorganisms as well as the development of resistance among old pathogens.[1,2] Thus, new sources of antimicrobial products have gained importance together with new approaches to synthesizing antimicrobial agents. Particularly, the molecule zwittermicin A has been discovered to display an antimicrobial potential.[3] It is produced by members of the *Bacillus cereus* group (especially *B. cereus* and *Bacillus thuringiensis*), which are abundantly and ubiquitously present in most soil communities.[4,5]

Zwittermicin A is a linear aminopolyol that was originally identified by virtue of its role in suppression of plant diseases by the *B. cereus* strain UW85.[3,6] It shows a broad spectrum of activity, as it may inhibit certain Gram-positive, Gram-negative (Table 1), and fungal organisms.[7] In general, zwittermicin A seems to be less inhibitory to Gram-positive than to Gram-negative bacteria; also, *B. cereus* strains that do not produce the antibiotic are generally susceptible to it. Interestingly, it also potentiates the insecticidal

*The Diverse Faces of Bacillus cereus*
ISBN 978-0-12-801474-5
http://dx.doi.org/10.1016/B978-0-12-801474-5.00010-4

117

**Table 1** Bacteria inhibited by zwittermicin A

| Gram-Negative bacteria | Gram-Positive bacteria |
| --- | --- |
| *Agrobacterium tumefaciens* A759 | *Bacillus megaterium* |
| *Bradyrhizobium japonicum* USDA 110 | *Bacillus cereus* 569 |
| *Cytophaga johnsonae* 9408 | *B. cereus* UW85 |
| *Erwinia carotovora* 8064 | *B. cereus* BAR145 |
| *Erwinia herbicola* IRQ | *B. cereus* SN14 |
| *E. herbicola* LS005 | *Bacillus subtilis* 168 |
| *Escherichia coli* K37 | *Bacillus thuringiensis* 4A9 |
| *Klebsiella pneumoniae* 8030 | *B. thuringiensis* 4D6 |
| *Pseudomonas aeruginosa* 9020 | *Clostridium pasteurianum* 5002 |
| *Pseudomonas fluorescens* 9023 | *Lactobacillus acidophilus* 4003 |
| *Rhizobium meliloti* 1021 | *Staphylococcus aureus* 3001 |
| *Rhizobium tropici* CIAT 899 | *Streptomyces griseus* 6501 |
| *Rhodobacter sphaeroides* 9502 | |
| *Rhodospirillum rubrum* 9405 | |
| *Salmonella typhimurium* LT2 | |
| *Vibrio cholerae* F115A | |
| *Yersinia pseudotuberculosis* | |

activity of the *B. thuringiensis* protein toxin, thus enhancing the mortality of insects that are typically refractory to killing, including gypsy moths reared on willow leaves.[8,9]

Several *B. cereus* and *B. thuringiensis* strains collected from plant roots and soils may synthesize zwittermicin A, as said,[4,6,8,10–13] and this compound is produced by strains from various geographical origins and cultivated from soils with different physical and biological properties.[8,14–16]

The mentioned *B. cereus* strain UW85 has been known to accumulate two antibiotics, that is, both zwittermicin A and kanosamine, in supernatants from cultures. As said, zwittermicin A is a novel, linear aminopolyol representing a new class of antibiotics.[4–6,10–13,17–44] It contributes to alfalfa damping-off suppression exerted by strain UW85 and may support other biological activities of this strain, such as control of fruit rot in cucumbers or suppression of other plant diseases.[10,12,13,35,42]

Zwittermicin A inhibits several bacteria, protists, fungi, and oomycetes as well; its activity is greater at higher pH than at lower; again, it is synergistic with the activity of kanosamine in inhibiting *Escherichia coli* and works additively with kanosamine to face the oomycete *Phytophthora*. The wide target range of this novel antibiotic suggests that zwittermicin A-producing bacteria, such as *B. cereus* UW85, might show utility for control of several foliar and soil-borne plant diseases.[4–6,10–13,17–44]

Also, zwittermicin A inhibitory activity observed against the lower eukaryotes (oomycetes, chrysophytes) suggests that this compound may deserve interest as to its potential to face the protist pathogens of humans as well, such as *Trichomonas* and *Giardia*.[22,37]

Zwittermicin A shares a similar target range, along with certain structural features, with chitosan, the latter being the deacetylated form of chitin and the polymeric form of glucosamine, and both chitosan and zwittermicin A are polycations. They both display greatest activity against the oomycetes,[19] both exert antibacterial properties, and both,[41] finally, can produce phytotoxic effects when present at high concentrations.[38,44] The polycationic nature of chitosan supports its biological activities, which include disruption of cell walls, inhibition of fungal RNA synthesis, DNA binding, and induction of a host resistance response in plants; conversely, authors who investigated zwittermicin A cannot thus far make any predictions about the target range or mode of action behind its activity.[26,27,29,41]

The discovery of antibiotic-producing microorganisms contributes to dealing with the challenges that confront agriculture and medicine. To improve human health, new drugs will be necessary to face the major human pathogens that have developed and will develop resistance to the antibiotic drugs that controlled them in past decades.[28] Likewise, to maintain food quality, improved measures for control of crop diseases need to be introduced, to replace fungicides that are currently and widely used; these chemical compounds are in fact likely to be restricted in the future owing to safety concerns along with the observation of pathogens' resistance to them.[21,23]

## BIOCONTROL IN ORGANISMS OTHER THAN *BACILLUS*

Production of compounds with antimicrobial activity is not only found in *Bacillus*, as diverse microorganisms may naturally compete with others in the ecological niches they inhabit. As an example, yeasts belonging to the genus *Metschnikowia* may release pulcherrimin, an antibiotic substance that is able to inhibit several bacterial and fungal species and that is currently used, therefore, to control postharvest pathogens. Pulcherrimin, specifically, is a reddish pigment that is able to chelate iron ions, thus immobilizing the metal in the medium where antagonized organisms cannot therefore grow; mutants that do not produce the pigment do not show the antimicrobial activity.[45–68] The latter may affect several bacterial and fungal species, including *Aspergillus flavus*, *Aspergillus versicolor*, *Aspergillus nidulans*,

*Aspergillus ustus, Aspergillus fumigatus, Aspergillus rubrum, Aspergillus terreus, Aspergillus flavipes, Trichophyton mentagrophytes, Trichophyton schoenleinii, Trichophyton violaceum,* and *Trichophyton tonsurans.*[45–68]

Likewise, some *Candida guilliermondii* strains exhibit a relevant potential in postharvest biocontrol of spoilage fungi during storage of vegetables and fruit. Particularly, the organism has been used to protect apples from post-harvest fruit-rotting fungi (that is, *Botrytis cinerea* and *Penicillium expansum*) and to prevent postharvest diseases caused by several other fungi (*P. expansum, Penicillium italicum, Penicillium digitatum, B. cinerea, Colletotrichum capsici, Rhizopus stolonifer, Rhizopus nigricans,* and *Botryodiplodia theobromae*). In addition, *C. guilliermondii* is lethal on the tomato root-knot nematode *Meloidogyne incognita.*[69]

While *B. cereus* biocontrol properties rely on zwittermicin A, multiple modes of action support the biocontrol activity of *C. guilliermondii*. The yeast may in fact attach to the hyphae of molds or compete for nutrients (glucose, sucrose, or nitrates), but does not seem to produce killer toxins.[69]

Antimicrobial compounds are also produced by bacteria other than *B. cereus*; they are named "bacteriocins" and are peptides or proteins that potentially kill closely related strains at very low concentrations. Nevertheless, three general features, their formation in ribosomes, their activity at low concentrations (in the nanomolar range), and their narrow spectrum of action, permit one to differentiate bacteriocins from antibiotics. Antibiotics are in fact synthesized by multienzymatic synthetase complexes, are usually effective at higher concentrations (in the micromolar range), and affect a broad spectrum of microorganisms. It is estimated that 99% of bacterial strains produce at least one bacteriocin; mainly owing to their narrow spectrum of activity, these compounds are not, however, used in agriculture, but participate only in growth regulation and competition against microbial communities in their respective habitats.[70–117]

Biological control activity in yeasts has notably emerged, instead, as mentioned above, as a safe alternative to fungicides to prevent and face decay loss in harvested commodities.[69]

In light of this, it is clear that much research is warranted to look for novel antimicrobial molecules aiming at controlling infections in plants and animals. Management of crop health is in fact essential to agriculture. Bio-control, that is, the use of microorganisms to inhibit plant pests, offers in this ambit an attractive option to chemical pesticides.[8]

Not only fungi, however, but numerous microorganisms have been identified that suppress plant pathologies both in the laboratory and in field experiments, but only few of these agents are effective over a broad range of conditions. This may be explained by the fact that each potential biocontrol strain evolved to compete in the habitat from which it is obtained and is not as well adapted to other environments.[8] Enhancing the effectiveness of bio-control in diverse natural niches is therefore essential to including it in standard agricultural practices.[8]

The majority of microorganisms outside fungi that have been investi-gated as potential biological agents are Gram-negative bacteria including *Agrobacterium*, Pseudomonadaceae, and *Erwinia*; finally, several *Bacillus* strains and actinomycetes have been described to exert this natural anti-microbial activity, but in general Gram-positive organisms have received far less intensive study than Gram-negative bacteria.[6] *Bacillus* species, however (like actinomycetes), show several characteristics that make them interesting candidates as biocontrol agents; mostly, they are abundant in soil and produce various biologically active metabolites, including zwit-termicin A.[6]

## CONCLUDING REMARKS

Entomologists have recognized and used diverse isolates of entomopatho-genic species that have potential biocontrol properties. Among these, a notable example is *B. thuringiensis*. Initially, this organism was acknowledged as an insect killer, but early field applications led to limited and variable results.[8] The subsequent development of methods to characterize new *B. thuringiensis* strains resulted in a vast array of inoculum strains that have extended the target range and enhanced the effectiveness of this species as a biocontrol organism.[8]

By virtue of identification of thousands of *B. thuringiensis* strains and of the understanding of the mechanism behind the insect control, *B. thuringiensis*-based products have gained increasing success over the years.[8]

Likewise, *B. cereus* UW85 (ATCC 53522) has been known to protect alfalfa seedlings from damping-off (a disease due to *Phytophthora medicaginis*), peanuts from *Sclerotinia minor*, tobacco seedlings from *Phytophthora nicotianae*, and cucumber fruits from *Pythium aphanidermatum*-related rot.[8]

In this field, therefore, there is the need for a deeper understanding of the biology of *B. cereus* and zwittermicin A production. This might have an increasing utility in biotechnology as well as in discovering other members of the zwittermicin A class displaying antibiotic activity that may have a place as valid alternatives to chemical pesticides, for a safer agriculture.

# REFERENCES

1. Broderick NA, Goodman RM, Handelsman J, Raffa KF. Effect of host diet and insect source on synergy of gypsy moth (Lepidoptera: Lymantriidae) mortality to *Bacillus thuringiensis* subsp. *kurstaki* by zwittermicin A. *Environ Entomol* 2003;**32**:387–91.
2. Broderick NA, Goodman RM, Raffa KF, Handelsman J. Synergy between zwittermicin A and *Bacillus thuringiensis* subsp. *kurstaki* against gypsy moth (Lepidoptera: Lymantriidae). *Environ Entomol* 2000;**29**:101–7.
3. Gold HS, Moellering Jr RC. Antimicrobial-drug resistance. *N Engl J Med* 1996;**335**: 1445–53.
4. He H, Silo-Suh LA, Clardy J, Handelsman J. Zwittermicin A, an antifungal and plant protection agent from *Bacillus cereus*. *Tetrahedron Lett* 1994;**35**:2499–502.
5. Raffel SJ, Stabb EV, Milner JL, Handelsman J. Genotypic and phenotypic analysis of zwittermicin A-producing strains of *Bacillus cereus*. *Microbiology* 1996;**42**:3425–36.
6. Silo-Suh LA, Lethbridge BJ, Raffel SJ, He H, Clardy J, Handelsman J. Biological activities of two fungistatic antibiotics produced by *Bacillus cereus* UW85. *Appl Environ Microbiol* 1994;**60**:2023–30.
7. Silo-Suh LA, Stabb EV, Raffel SJ, Handelsman J. Target range of zwittermicin A, an aminopolyol antibiotic from *Bacillus cereus*. *Curr Microbiol* 1998;**37**:6–11.
8. Stabb EV, Jacobson LM, Handelsman J. Zwittermicin A-producing strains of *Bacillus cereus* from diverse soils. *Appl Environ Microbiol* 1994;**60**:4404–12.
9. Virk A, Steckelberg JM. Clinical aspects of antimicrobial resistance. *Mayo Clin Proc* 2000;**75**:200–14.
10. Handelsman J, Raffel S, Mester EH, Wunderlich L, Grau CR. Biological control of damping-off of alfalfa seedlings with *Bacillus cereus* UW85. *Appl Environ Microbiol* 1990;**56**:713–8.
11. Handelsman J, Nesmith WS, Raffel SJ. Microassay for biological and chemical control of infection of tobacco by *Phytophthora parasitica* var. *nicotianae*. *Curr Microbiol* 1991;**22**:317–9.
12. Smith KP, Havey MJ, Handelsman J. Suppression of cottony leak of cucumber with *Bacillus cereus* strain UW85. *Plant Dis* 1993;**77**:139–42.
13. Osburn RM, Milner JL, Oplinger ES, Smith RS, Handelsman J. Effect of *Bacillus cereus* UW85 on the yield of soybean at two field sites in Wisconsin. *Plant Dis* 1995;**79**: 551–6.
14. DeLong ER, Wickham GS, Pace NR. Phylogenetic strains: ribosomal RNA-based probes for the identification of single cells. *Science* 1989;**243**:1360–3.
15. Hahn D, Amann RI, Ludwig W, Akkermans ADL, Schleifer KH. Detection of microorganisms in soil after *in situ* hybridization with rRNA-targeted, fluorescently labelled oligonucleotides. *J Gen Microbiol* 1992;**38**:879–87.
16. Assmus B, Hutzler P, Kirchhof G, Amann R, Lawrence JR, Hartmann A. In situ localization of *Apospirillum brasdense* in the rhizosphere of wheat with fluorescently labeled, rRNA targeted oligonucleotide probes and scanning confocal laser microscopy. *Appl Environ Microbiol* 1995;**61**:1013–9.
17. Agrios GN. Plant diseases caused by fungi. In: *Plant pathology*. San Diego (California): Academic Press Inc.; 1997. p. 245–406.
18. Alexopolous CJ, Mims CW, Blackwell M. *Introductory mycology*. 4th ed. New York: John Wiley & Sons Inc.; 1996.
19. Allan CR, Hadwiger LA. The fungicidal effect of chitosan on fungi of varying cell wall composition. *Exp Mycol* 1979;**3**:285–7.
20. Beale AS, Sutherland R. Measurement of combined antibiotic action. In: Russel AD, Quesnel LB, editors. *Antibiotics: assessment of antimicrobial activity and resistance*. New York (NY): Academic Press; 1983. p. 299–315.

21. Brent KJ. Fungicide resistance in crops—its practical significance and management. In: Brent KJ, Atkin RK, editors. *Rational pesticide use*. Great Britain: Cambridge University Press; 1987. p. 137–51.
22. Brooks DR. The importance of protistan phylogeny for macroevolution. *Biosystems* 1988;**21**:189–96.
23. Dekker J. Build-up and persistence of fungicide resistance. In: Brent KJ, Atkin RK, editors. *Rational pesticide use*. Great Britain: Cambridge University Press; 1987. p. 153–68.
24. Dhingra OD, Sinclair JB. *Basic plant pathology methods*. Boca Raton (FL): CRC Press Inc.; 1985. p. 308.
25. Gunderson JH, Elwood H, Ingold A, Kindle K, Sogin ML. Phylogenetic relationships between chlorophytes, chrysophytes, and oomycetes. *Proc Natl Acad Sci USA* 1987;**84**:5823–7.
26. Hadwiger LA, Beckman JM. Chitosan as a component of pea-*Fusarium solani* interactions. *Plant Physiol* 1980;**66**:205–11.
27. Hadwiger LA, Kendra DF, Fristensky BW, Wagoner W. Chitosan both activates genes in plants and inhibits RNA synthesis in fungi. In: Muzzarelli R, Jeuniaux C, Gooday GW, editors. *Chitin in nature and technology*. New York (NY): Plenum Press; 1986. p. 209–14.
28. Lederberg JR, Shope E, Oaks SC. Addressing the threats. In: *Emerging infections. Microbial threats to health in the United States*. Washington (DC): National Academy Press; 1992. p. 113–91.
29. Leuba JL, Stossel P. Chitosan and other polyamines: antifungal activity and interaction with biological membranes. In: Muzzarelli R, Jeuniaux C, Gooday GW, editors. *Chitin in nature and technology*. New York (NY): Plenum Press; 1986. p. 215–22.
30. Maniatis T, Fritsch EF, Sambrook J. *Molecular cloning, a laboratory manual*. Cold Spring Harbor (NY): Cold Spring Harbor Laboratory Press; 1982. p. 440.
31. Milner JL, Raffel SJ, Lethbridge BJ, Handelsman J. Culture conditions that influence accumulation of zwittermicin A by *Bacillus cereus* UW85. *Appl Microbiol Biotechnol* 1995;**43**:685–91.
32. Milner JL, Silo-Suh L, Lee JC, He H, Clardy J, Handelsman J. Production of kanosamine by *Bacillus cereus* UW85. *Appl Environ Microbiol* 1996;**62**:3061–5.
33. Milner JL, Stohl EA, Handelsman J. Zwittermicin A resistance gene from *Bacillus cereus*. *J Bacteriol* 1996;**178**:4266–72.
34. Mitchell JE, Yang CY. Factors affecting growth and development of *Aphanomyces euteiches*. *Phytopathology* 1966;**56**:917–22.
35. Phipps PM. Evaluation of biological agents for control of sclerotinia blight of peanut, 1991. *Biol Cult Tests Control Plant Dis* 1992;**7**:60.
36. Rahimian MK, Banihashemi Z. A method for obtaining zoospores of *Pythium aphanidermatum* and their use in determining cucurbit seedling resistance to damping-off. *Plant Dis Rep* 1979;**63**:658–61.
37. Roger AJ, Clark CG, Doolittle WF. A possible mitochondrial gene in the early-branching amitochondriate protist *Trichomonas vaginalis*. *Proc Natl Acad Sci USA* 1996;**93**:14618–22.
38. Silo-Suh LA. *Biological activities of two antibiotics produced by Bacillus cereus UW85* [Ph.D. thesis]. University of Wisconsin-Madison; 1994.
39. Stabb EV, Handelsman J. Genetic analysis of zwittermicin A resistance in *Escherichia coli*: effects on membrane potential and RNA polymerase. *Mol Microbiol* 1998;**27**:311–22.
40. Stohl EA, Stabb EV, Handelsman J. Zwittermicin A and biological control of oomycete pathogens. In: Stacey G, Mullin B, Gresshoff PM, editors. *Biology of plant-microbe interactions*. St. Paul (MN): Int Soc Mol Plant-Microbe Interact; 1996. p. 475–80.

41. Sudarshan NR, Hoover DG, Knorr D. Antibacterial action of chitosan. *Food Biotechnol* 1992;**6**:257–72.
42. Thiemann JE, Beretta G. Antiprotozoal antibiotics. *J Antibiot Series A (Tokyo)* 1967;**20**: 191–3.
43. Tuite J. *Plant pathological methods*. Minneapolis (MN): Burgess Publishing; 1969. p. 112.
44. Young DH, Kauss H. Release of calcium from suspension cultured *Glycine max* cells by chitosan, other polycations, and polyamines in relation to effects on membrane permeability. *Plant Physiol* 1983;**73**:698–702.
45. Sisti M, Savini V. Antifungal properties of the human *Metschnikowia* strain IHEM 25107. *Folia Microbiol* 2014;**59**(3):263–6.
46. Savini V, Hendrickx M, Sisti M, Masciarelli G, Favaro M, Fontana C, et al. An atypical, pigment-producing *Metschnikowia* strain from a leukaemia patient. *Med Mycol* 2013;**51**(4):438–43.
47. Fidalgo-Jiménez A, Daniel HM, Evrard P, Decock C, Lachance MA. *Metschnikowia cubensis* sp. nov., a yeast species isolated from flowers in Cuba. *Int J Syst Evol Microbiol* 2008;**58**:2955–61.
48. Molnár O, Prillinger H. Analysis of yeast isolates related to *Metschnikowia pulcherrima* using the partial sequences of the large subunit rDNA and the actin gene; description of *Metschnikowia andauensis* sp. nov. *Syst Appl Microbiol* 2005;**28**:717–26.
49. Nguyen NH, Suh SO, Erbil CK, Blackwell M. *Metschnikowia noctiluminum* sp. nov., *Metschnikowia corniflorae* sp. nov., and *Candida chrysomelidarum* sp. nov., isolated from green lacewings and beetles. *Mycol Res* 2006;**110**:346–56.
50. Lachance MA, Ewing CP, Bowles JM, Starmer WT. *Metschnikowia hamakuensis* sp. nov., *Metschnikowia kamakouana* sp. nov. and *Metschnikowia mauinuiana* sp. nov., three endemic yeasts from Hawaiian nitidulid beetles. *Int J Syst Evol Microbiol* 2005;**55**:1369–77.
51. Lachance MA, Bowles JM, Wiens F, Dobson J, Ewing CP. *Metschnikowia orientalis* sp. nov., an Australasian yeast from nitidulid beetles. *Int J Syst Evol Microbiol* 2006;**56**: 2489–93.
52. Lachance MA, Bowles JM. *Metschnikowia arizonensis* and *Metschnikowia dekortorum*, two new large-spored yeast species associated with floricolous beetles. *FEMS Yeast Res* 2002;**2**:81–6.
53. Xue ML, Zhang LQ, Wang QM, Zhang JS, Bai FY. *Metschnikowia sinensis* sp. nov., *Metschnikowia zizyphicola* sp. nov. and *Metschnikowia shanxiensis* sp. nov., novel yeast species from jujube fruit. *Int J Syst Evol Microbiol* 2006;**56**:2245–50.
54. Hong SG, Bae KS, Herzberg M, Titze A, Lachance MA. *Candida kunwiensis* sp. nov., a yeast associated with flowers and bumblebees. *Int J Syst Evol Microbiol* 2003;**53**:367–72.
55. Giménez-Jurado G, Valderrama MJ, Sá-Nogueira I, Spencer-Martins I. Assessment of phenotypic and genetic diversity in the yeast genus *Metschnikowia*. *Antonie Van Leeuwenhoek* 1995;**68**:101–10.
56. Pospíšil L. The significance of *Candida pulcherrima* findings in human clinical specimens. *Mycoses* 1989;**32**:581–3.
57. Suh SO, Gibson CM, Blackwell M. *Metschnikowia chrysoperlae* sp. nov, *Candida picachoensis* sp. nov. and *Candida pimensis* sp. nov., isolated from the green lacewings *Chrysoperla comanche* and *Chrysoperla carnea* (Neuroptera: Chrysopidae). *Int J Syst Evol Microbiol* 2004;**54**:1883–90.
58. Guerzoni E, Marchetti R. Analysis of yeast flora associated with grape sour rot and of the chemical disease markers. *Appl Environ Microbiol* 1987;**53**:571–6.
59. Bleve G, Grieco F, Cozzi G, Logrieco A, Visconti A. Isolation of epiphytic yeasts with potential for biocontrol of *Aspergillus carbonarius* and *A. niger* on grape. *Int J Food Microbiol* 2006;**108**:204–9.
60. Combina M, Elía A, Mercado L, Catania C, Ganga A, Martinez C. Dynamics of indigenous yeast populations during spontaneous fermentation of wines from Mendoza, Argentina. *Int J Food Microbiol* 2005;**99**:237–43.

61. Cocolin L, Urso R, Rantsiou K, Cantoni C, Comi G. Dynamics and characterization of yeasts during natural fermentation of Italian sausages. *FEMS Yeast Res* 2006;**6**: 692–701.

62. Dorko E, Kmetová M, Pilipcinec E, Bracoková I, Dorko F, Danko J, et al. Rare non-albicans *Candida* species detected in different clinical diagnoses. *Folia Microbiol* 2000;**45**:364–8.

63. Jautová J, Virágová S, Ondrasovic M, Holoda E. Incidence of *Candida* species isolated from human skin and nails: a survey. *Folia Microbiol* 2001;**46**:333–7.

64. Kluyver AJ, van der Walt JP, van Triet AJ. Pulcherrimin, the pigment of *Candida pulcherrima*. *Proc Natl Acad Sci USA* 1953;**39**:583–93.

65. Türkel S, Ener B. Isolation and characterization of new *Metschnikowia pulcherrima* strains as producers of the antimicrobial pigment pulcherrimin. *Z Naturforsch C* 2009;**64**:405–10.

66. Lachance MA. The yeasts, a taxonomic study. In: 5th ed. Kurtzman CP, Fell JW, Boekhout T, editors. *Metschnikowia kamienski*, vol. 1899. Amsterdam: Elsevier; 2011. p. 575–620.

67. Uffen RL, Canale-Paola E. Synthesis of pulcherrimic acid by *Bacillus subtilis*. *J Bacteriol* 1972;**111**(86):93.

68. Sipiczki M. *Metschnikowia* strains isolated from botrytized grapes antagonize fungal and bacterial growth by iron depletion. *Appl Environ Microbiol* 2006;**72**:6716–24.

69. Papon N, Savini V, Lanoue A, Simkin AJ, Crèche J, Giglioli-Guivarc'h N, et al. *Candida guilliermondii*: biotechnological applications, perspectives for biological control, emerging clinical importance and recent advances in genetics. *Curr Genet* 2013;**59**(3):73–90.

70. Klaenhammer TR. Genetics of bacteriocins produced by lactic acid bacteria. *FEMS Microbiol Rev* 1993;**12**:39–85.

71. Hammami R, Zouhir A, Le Lay C, Ben Hamida J, Fliss I. BACTIBASE second release: a database and tool platform for bacteriocin characterization. *BMC Microbiol* 2010;**10**:22.

72. Riley MA, Chavan MA, editors. *Bacteriocins ecology and evolution*. Berlin: Springer; 2007.

73. Drider D, Rebuffat S, editors. *Prokaryotic antimicrobial peptides*. New York: Springer; 2011.

74. Smarda J, Benada O. Phage tail-like (high-molecular-weight) bacteriocins of *Budvicia aquatica* and *Pragia fontium* (Enterobacteriaceae). *Appl Environ Microbiol* 2005;**71**: 8970–3.

75. Gebhart D, Williams SR, Bishop-Lilly KA, Govoni GR, Willner KM, Butani A, et al. Novel high-molecular-weight, R-type bacteriocins of *Clostridium difficile*. *J Bacteriol* 2012;**194**:6240–7.

76. Michel-Briand Y, Baysse C. The pyocins of *Pseudomonas aeruginosa*. *Biochimie* 2002;**84**:499–510.

77. O'Connor EB, Cotter PD, O'Connor P, O'Sullivan O, Tagg JR, Ross RP, et al. Relatedness between the two-component lantibiotics lacticin 3147 and staphylococcin C55 based on structure, genetics and biological activity. *BMC Microbiol* 2007;**7**:24.

78. Cotter PD, Hill C, Ross RP. Bacteriocins: developing innate immunity for food. *Nat Rev Microbiol* 2005;**3**:777–88.

79. Havarstein LS, Diep DB, Nes IF. A family of bacteriocin ABC transporters carry out proteolytic processing of their substrates concomitant with export. *Mol Microbiol* 1995;**16**:229–40.

80. Venema K, Kok J, Marugg JD, Toonen MY, Ledeboer AM, Venema G, et al. Functional analysis of the pediocin operon of *Pediococcus acidilactici* PAC1.0: PedB is the immunity protein and PedD is the precursor processing enzyme. *Mol Microbiol* 1995;**17**:515–22.

81. Nes IF, Eijsink VGH. Regulation of group II peptide bacteriocin synthesis by quorum sensing mechanisms. In: Dunny GM, Winans SC, editors. *Cell–cell signalling in bacteria*. Washington: ASM American Society for Microbiology; 1999.

82. Gillor O, Vriezen JA, Riley MA. The role of SOS boxes in enteric bacteriocin regulation. *Microbiology* 2008;**154**:1783–92.

83. Inglis RF, Bayramoglu B, Gillor O, Ackermann M. The role of bacteriocins as selfish genetic elements. *Biol Lett* 2013;**9**. 20121173.

84. Van Melderen L, Saavedra De Bast M. Bacterial toxin-antitoxin systems: more than selfish entities? *PLoS Genet* 2009;**5**:e1000437.

85. Wladyka B, Wielebska K, Wloka M, Bochenska O, Dubin G, Dubin A, et al. Isolation, biochemical characterization, and cloning of a bacteriocin from the poultry-associated *Staphylococcus aureus* strain CH-91. *Appl Microbiol Biotechnol* 2013;**97**:7229–39.

86. Islam MR, Nishie M, Nagao J, Zendo T, Keller S, Nakayama J, et al. Ring A of nukacin ISK-1: a lipid II-binding motif for type-A(II) lantibiotic. *J Am Chem Soc* 2012;**134**: 3687–90.

87. Minamikawa M, Kawai Y, Inoue N, Yamazaki K. Purification and characterization of warnericin RB4, anti-*Alicyclobacillus* bacteriocin, produced by *Staphylococcus warneri* RB4. *Curr Microbiol* 2005;**51**:22–6.

88. Ceotto H, Holo H, da Costa KF, Nascimento Jdos S, Salehian Z, Nes IF, et al. Nukacin 3299, a lantibiotic produced by *Staphylococcus simulans* 3299 identical to nukacin ISK-1. *Vet Microbiol* 2010;**146**:124–31.

89. Bierbaum G, Sahl HG. Autolytic system of *Staphylococcus simulans* 22: influence of cationic peptides on activity of N-acetylmuramoyl-ʟ-alanine amidase. *J Bacteriol* 1987;**169**:5452–8.

90. Bonelli RR, Schneider T, Sahl HG, Wiedemann I. Insights into in vivo activities of lantibiotics from gallidermin and epidermin mode-of-action studies. *Antimicrob Agents Chemother* 2006;**50**:1449–57.

91. Netz DJ, Pohl R, Beck-Sickinger AG, Selmer T, Pierik AJ, Bastos Mdo C, et al. Biochemical characterisation and genetic analysis of aureocin A53, a new, atypical bacteriocin from *Staphylococcus aureus*. *J Mol Biol* 2002;**319**:745–56.

92. Sandiford S, Upton M. Identification, characterization, and recombinant expression of epidermicin NI01, a novel unmodified bacteriocin produced by *Staphylococcus epidermidis* that displays potent activity against staphylococci. *Antimicrob Agents Chemother* 2012;**56**:1539–47.

93. Sawa N, Koga S, Okamura K, Ishibashi N, Zendo T, Sonomoto K. Identification and characterization of novel multiple bacteriocins produced by *Lactobacillus sakei* D98. *J Appl Microbiol* 2013;**115**:61–9.

94. Yoneyama F, Ohno K, Imura Y, Li M, Zendo T, Nakayama J, et al. Lacticin Q-mediated selective toxicity depending on physicochemical features of membrane components. *Antimicrob Agents Chemother* 2011;**55**:2446–50.

95. Navaratna MA, Sahl HG, Tagg JR. Two-component anti-*Staphylococcus aureus* lantibiotic activity produced by *Staphylococcus aureus* C55. *Appl Environ Microbiol* 1998;**64**:4803–8.

96. Netz DJ, Sahl HG, Marcelino R, dos Santos Nascimento J, de Oliveira SS, Soares MB, et al. Molecular characterisation of aureocin A70, a multi-peptide bacteriocin isolated from *Staphylococcus aureus*. *J Mol Biol* 2001;**311**:939–49.

97. Navaratna MA, Sahl HG, Tagg JR. Identification of genes encoding two-component lantibiotic production in *Staphylococcus aureus* C55 and other phage group II *S. aureus* strains and demonstration of an association with the exfoliative toxin B gene. *Infect Immun* 1999;**67**:4268–71.

98. Garneau S, Martin NI, Vederas JC. Two-peptide bacteriocins produced by lactic acid bacteria. *Biochimie* 2002;**84**:577–92.

99. Varella Coelho ML, Santos Nascimento JD, Fagundes PC, Madureira DJ, Oliveira SS, Vasconcelos de Paiva Brito MA, et al. Activity of staphylococcal bacteriocins against *Staphylococcus aureus* and *Streptococcus agalactiae* involved in bovine mastitis. *Res Microbiol* 2007;**158**:625–30.

100. Schindler CA, Schuhardt VT. Lysostaphin: a new bacteriolytic agent for the *Staphylococcus*. *Proc Natl Acad Sci USA* 1964;**51**:414–21.
101. Gagliano VJ, Hinsdill RD. Characterization of a *Staphylococcus aureus* bacteriocin. *J Bacteriol* 1970;**104**:117–25.
102. Nakamura T, Yamazaki N, Taniguchi H, Fujimura S. Production, purification, and properties of a bacteriocin from *Staphylococcus aureus* isolated from saliva. *Infect Immun* 1983;**39**:609–14.
103. Hale EM, Hinsdill RD. Characterization of a bacteriocin from *Staphylococcus aureus* strain 462. *Antimicrob Agents Chemother* 1973;**4**:634–40.
104. Crupper SS, Iandolo JJ. Purification and partial characterization of a novel antibacterial agent (Bac1829) produced by *Staphylococcus aureus* KSI1829. *Appl Environ Microbiol* 1996;**62**:3171–5.
105. Iqbal A, Ahmed S, Ali SA, Rasool SA. Isolation and partial characterization of Bac201: a plasmid-associated bacteriocin-like inhibitory substance from *Staphylococcus aureus* AB201. *J Basic Microbiol* 1999;**39**:325–36.
106. Saeed S, Ahmad S, Rasool SA. Antimicrobial spectrum, production and mode of action of staphylococcin 188 produced by *Staphylococcus aureus* 188. *Pak J Pharm Sci* 2004;**17**: 1–8.
107. Scott JC, Sahl HG, Carne A, Tagg JR. Lantibiotic-mediated anti-lactobacillus activity of a vaginal *Staphylococcus aureus* isolate. *FEMS Microbiol Lett* 1992;**72**:97–102.
108. Daly KM, Upton M, Sandiford SK, Draper LA, Wescombe PA, Jack RW, et al. Production of the Bsa lantibiotic by community-acquired *Staphylococcus aureus* strains. *J Bacteriol* 2010;**192**:1131–42.
109. Saeed S, Rasool SA, Ahmed S, Khanum T, Khan MB, Abbasi A, et al. New insight in staphylococcin research: bacteriocin and/or bacteriocin-like inhibitory substance(s) produced by *Stapphylococcus aureus* AB188. *World J Microbiol Biotech* 2006;**22**:713–22.
110. Marshall BM, Levy SB. Food animals and antimicrobials: impacts on human health. *Clin Microbiol Rev* 2011;**24**:718–33.
111. Harrison EF, Cropp CB. Comparative in vitro activities of lysostaphin and other antistaphylococcal antibiotics on clinical isolates of *Staphylococcus aureus*. *Appl Microbiol* 1965;**13**:212–5.
112. von Eiff C, Kokai-Kun JF, Becker K, Peters G. In vitro activity of recombinant lysostaphin against *Staphylococcus aureus* isolates from anterior nares and blood. *Antimicrob Agents Chemother* 2003;**47**:3613–5.
113. Climo MW, Patron RL, Goldstein BP, Archer GL. Lysostaphin treatment of experimental methicillin-resistant *Staphylococcus aureus* aortic valve endocarditis. *Antimicrob Agents Chemother* 1998;**42**:1355–60.
114. Oluola O, Kong L, Fein M, Weisman LE. Lysostaphin in treatment of neonatal *Staphylococcus aureus* infection. *Antimicrob Agents Chemother* 2007;**51**:2198–200.
115. Senturk S, Cetin C, Temizel M, Ozel E. Evaluation of the clinical efficacy of subconjunctival injection of clindamycin in the treatment of naturally occurring infectious bovine keratoconjunctivitis. *Vet Ophthalmol* 2007;**10**:186–9.
116. Saeed S, Rasool SA, Ahmad S, Zaidi SZ, Rehmani S. Antiviral activity of staphylococcin 188: a purified bacteriocin-like inhibitory substance isolated from *Staphylococcus aureus* AB188. *Res J Microbiol* 2007;**2**:796–806.
117. Savini V, Catavitello C, Bianco A, Balbinot A, D'Antonio D. Epidemiology, pathogenicity and emerging resistances in *Staphylococcus pasteuri*: from mammals and lampreys, to man. *Recent Pat Antiinfect Drug Discov* 2009;**4**:123–9.

# CHAPTER 11

# *Bacillus* Species Outside the *Bacillus cereus* Group

**Vincenzo Savini**
Clinical Microbiology and Virology, Laboratory of Bacteriology and Mycology, Civic Hospital of Pescara, Pescara, Italy

## SUMMARY

Although *Bacillus cereus* is the most important agent of enteric disease within the genus, other *Bacillus* species outside the *B. cereus* group are toxigenic. In fact, *Bacillus licheniformis, Bacillus subtilis, Bacillus mojavensis, Bacillus pumilus,* and *Bacillus fusiformis* may be toxin producers, and, particularly, strains of these species can produce cytotoxins and emetic toxins, but not *B. cereus*-like enterotoxins.

## ENTEROTOXINS

The genus *Bacillus* includes a heterogenic group of endospore-forming, facultative rods that are widely distributed in nature. Spore formation allows them to better tolerate adverse conditions compared to non-sporulating enteropathogenic bacteria and to proliferate in a broad range of environments such as water and processed and untreated food products.[1]

*Bacillus cereus* is the major cause of food poisoning within this genus by virtue of enterotoxins and the emetic toxin it may produce,[2] whereas other *Bacillus* species have commonly been considered of poor clinical significance in the context of food poisoning events.[3] Nonetheless, their ability to produce both enterotoxins and the emetic toxin has been increasingly acknowledged, particularly in certain strains belonging to the species *Bacillus subtilis, Bacillus pumilus,* and *Bacillus licheniformis,* which have been linked to episodes of food-borne disease. *B. licheniformis*[4–6]

Whenever species identification of *Bacillus* is reached without performing DNA-based assays the results are often incomplete, and in earlier papers published in past years and related to *Bacillus*-associated food poisoning and

*The Diverse Faces of Bacillus cereus*
ISBN 978-0-12-801474-5
http://dx.doi.org/10.1016/B978-0-12-801474-5.00011-6

toxin identification (with the exception of *B. cereus*), the latter was based mainly on biochemical methods or else the identification methodologies were not explained at all.[1] As an example, *B. subtilis* (with its subspecies *B. subtilis* subsp. *subtilis* and *B. subtilis* subsp. *spizizenii*) and *Bacillus mojavensis* are very close genetically, and discrimination between them can be achieved only by using molecular methods, including DNA–DNA hybridization[7] or partial *gyrA* gene sequence analysis.[8] It emerges, then, that unreliable identification at the species level of *Bacillus* strains might have reasonably affected past investigations on species-related toxins.[1]

There is no doubt, however, that numerous *Bacillus* species other than those of the *B. cereus* group can be responsible for food poisoning.[4] Nevertheless, with the exception of an emetic toxin from *B. licheniformis*, toxins from these species have not been purified nor characterized.[4,6,9]

Several authors reported *B. cereus*-like enterotoxins from non-*B. cereus* species of the genus, but without providing a conclusive identification of those toxins. It should also be noted that the regulation of transcription of cytotoxins and enterotoxins in *B. cereus* strains is controlled by the positive regulator PlcR,[10] and if the enterotoxin genes were acquired (from *B. cereus*) by members belonging to other *Bacillus* species, they would need either a modification in the promoter (to be transcribed) or a cotransfer of the regulator gene; this is not likely to occur, however, as the two genes are placed on different genome sections.[1] Also, *Bacillus circulans*, *Bacillus megaterium*, *Bacillus lentimorbis*, *Bacillus amyloliquefaciens*, *Bacillus lentus*, and the above-mentioned *B. subtilis* and *B. licheniformis* have been known to carry the *hbl* and *nhe* genes or to produce the proteins they encode (as observed through commercially available kits) but, unless these species are identified by DNA analysis, their characterization may be debatable.[1]

Based on current knowledge, then, no other cytotoxin-producing *Bacillus* strains seem to share toxins with *B. cereus*.[1,3,11–13]

## EMETIC TOXINS

Ring-form toxins have been found in *B. cereus* (cereulide), as well as in *B. licheniformis* and *B. pumilus*.[1,14] These toxins seem to be part of the same family of ring-form peptides as cereulide.[1,14] Moreover, *B. mojavensis*, *B. subtilis*, and *B. pumilus* may produce peptides of this type that are distinctively different in size from cereulide and, notably, such putative emetic toxins described in these species are produced better or just as well at 22 °C as at 30 °C.[1,14] Conversely, *B. cereus* shows an optimal temperature of production that is much lower (12–22 °C), while *B. licheniformis* probably optimally produces the emetic toxin at close to 28 °C (this temperature was used for documenting toxin production when it was first characterized in *B. licheniformis*).[1,15–17] Whether putative emetic toxins are true emetic toxins, however, can be verified only by feeding experiments with primates.[1]

Nonetheless, *Bacillus* species other than those included in the *B. cereus* group seem to be involved in food poisoning cases that have been related to ingestion of foods such as pizza and pasta, although rarely.[1,22,23] Such data, however, highlight that knowledge of the enteric pathogenic potential and related virulence traits expressed by *Bacillus* strains outside the *B. cereus* group may still be far from being completely understood.

## FOODS

Toxin-producing *Bacillus* other than organisms of the *B. cereus* group seem to be rare, and none of the toxins described share similarity with the *B. cereus*-related enterotoxins and cereulide, which are responsible for the pathogenesis of most *Bacillus*-related enteritis and food poisoning episodes.[18–21] However, rarely, *Bacillus* species other than those of the *B. cereus* group might be involved in food poisoning cases through foods such as pizza and pasta (particularly when these are left overnight without being refrigerated).[1,22,23] Again, the variety of ready-to-eat, lightly heated food products that are thought to have a several-week shelf life if cooled properly could lead to food poisoning by virtue of surviving spores that will then germinate.[1] In light of this, it is hoped that the food industry would be able to detect possible virulent *Bacillus* species, even outside the *B. cereus* group, in food ingredients and food production plants. This detection, however, requires knowledge of the toxins produced as well as of the bacterial genes.[1]

## *BACILLUS LICHENIFORMIS*

Diverse *Bacillus* species other than those of the *B. cereus* group seem then to be related to food poisoning cases and can share production of heat-stable toxins (such as *B. cereus* cereulide); also, strains of the species *B. licheniformis*, *B. pumilus*, and *B. amyloliquefaciens* with a food poisoning origin have been observed to produce substances exerting toxic activity against mammalian cells; of course, this is a risk for the dairy industry as both endospores and toxins may tolerate milk product processing such as pasteurization (occurring at 74 °C for 15–20 s) and whey evaporation (50–70 °C).[24]

A source of toxigenic *Bacillus* in raw milk is represented by cows with mastitis[25–40]; strains from milk of these bovines should not be found in milk products since mastitic milk is not generally sent to dairies. Nevertheless, toxigenic *B. licheniformis* has been isolated from the udder of a bovine that appeared clinically to have recovered from the mammary gland infection.[24]

*B. licheniformis* inhabits the natural environment; it typically forms large, β-hemolytic, blister colonies that become opaque, wrinkled, and rough after additional incubation (see the picture of a *B. licheniformis* isolate from the Bacteriology and Mycology Laboratory, Civic Hospital of Pescara, Italy, on this book's cover, on the right side). The organism has been increasingly recognized as a human pathogen; in particular, it has been responsible for cases of bacteremia, food poisoning, ophthalmitis, ventriculitis, brain abscess, and a deep skin abscess (following a plant thorn injury).[41–46] Also, a bacteremic episode was found to be related to intentional injection of organic drain cleaner (containing spores of *B. licheniformis*, *B. subtilis*, and *Bacillus polymyxa*) bilaterally into the patient's antecubital veins and subcutaneous tissues as a suicide attempt. Finally, a number of septicemias were due to contamination of indwelling catheters or followed intravenous injection in drug abusers.[41–43] It is of particular interest that, as highlighted in the above-mentioned suicide attempt case, one should consider the pathogenic potential of numerous organic consumer products that have been labeled as harmless.[41] In this context, owing to their dormant spores and saprophytic nature, members of the genus *Bacillus* are manufactured to be used as drain and septic tank cleaners, as well as insecticides and pesticides. Both septic tank products and drains contain spores belonging to *B. licheniformis* as major species.[41] Likewise, *Bacillus thuringiensis* (which is, however, included in the *B. cereus* group) has been widely used as a biocontrol organism against insect pests in agriculture and forestry, as well as to control vectors. [41] In this ambit, it is of concern that a case of corneal ulcer subsequent to eye exposure to an insecticide containing *B. thuringiensis* has been described.[41–44] It would be important, therefore, that manufacturers might promptly provide information about organisms present in their products, in the event of a product-related disease, as well as data on antibiotic susceptibilities, upon request, if any, by treating clinicians.[41]

Again, in the prosthetic aortic valve endocarditis case described by Santini,[42] the patient had no known immune system impairment and, notably, when the partially detached artificial valve was removed, a limited abscess was observed underneath the aortic ring; interestingly, cultures of both the explanted prosthesis and the abscess yielded a pure growth of *B. licheniformis*.[42]

A peritonitis case involved instead a woman with a history of hypertension; she had been treated with hemodialysis initially and subsequently with continuous ambulatory peritoneal dialysis given the occlusion of the arteriovenous fistula. She was having the third episode of peritonitis when the dialysis effluent showed cloudiness, and a white blood cell count of 945 elements mm$^{-3}$, and the effluent culture grew *B. licheniformis*.[45]

Finally, in the veterinary setting, *B. licheniformis* can often be collected from the uterine lumen of cattle with and without metritis, and it has been documented that cattle from which the organism could be isolated showed greater acute-phase protein responses than animals from which the bacterium was not cultivated.[47]

*B. licheniformis* inhabits the natural environment, including marine sponge, foods, and wastewaters; it may display immunostimulatory activity in fishes; additionally, it has been used as a probiotic against diarrhea in humans and, finally, has long been considered as a harmless contaminant when isolated from clinical materials.[48–60] Nevertheless, its potential pathogenicity should be reconsidered, in light of the increasingly reported disease cases both in humans and in animals, including its enteric pathogenic potential and its propensity to cause disease in bodily compartments and tissues outside the gut. Furthermore, this species is of ecologic and agricultural interest as certain strains may exert biocontrol activity against phytopathogenic microorganisms. For instance, *B. licheniformis* strain S213 is effective against the fungus *Phoma medicaginis*, which causes dumping-off disease in *Medicago truncatula*; and again, *B. licheniformis* HS10 shows biocontrol properties against *Pseudoperonospora cubensis*, a fungus that is responsible for cucumber downy disease.[50,58]

To conclude, aside from *Bacillus anthracis*, *B. cereus* is still the major agent of disease (mostly food-borne intoxications) within the genus *Bacillus*[25,37,61–90]; nevertheless, identification to the species level of *Bacillus* organisms outside the *B. cereus* group, as well as purification and characterization of potential toxins they form, is gaining increasing interest in the medical setting.[1] However, much investigation is warranted in this field, as clarification of the roles these bacteria may play in the etiology and pathogenesis of human infectious diseases, including enteric illnesses, still is, at this time, a work in progress.

# REFERENCES

1. From C, Pukall R, Schumann P, Hormazábal V, Granum PE. Toxin-producing ability among *Bacillus* spp. outside the *Bacillus cereus* group. *Appl Environ Microbiol* 2005;**71**: 1178–83.
2. Granum PE. Food microbiology: fundamentals and frontiers. In: Doyle MP, editor. *Bacillus cereus*. Washington: ASM Press; 2001. p. 373–81.
3. Rowan NJ, Deans K, Anderson JG, Gemmel CG, Hunter IS, Chaithong T. Putative virulence factor expression by clinical and food isolates of *Bacillus* spp. after growth in reconstituted infant milk formulae. *Appl Environ Microbiol* 2001;**67**:3873–81.
4. Kramer JM, Gilbert RJ. Foodborne bacterial pathogens. In: Doyle MP, editor. *Bacillus cereus and other Bacillus species*. New York: Marcel Dekker; 1989. p. 22–70.
5. Pedersen PB, Bjørnvad ME, Rasmussen MD, Petersen JN. Cytotoxic potential of industrial strains of *Bacillus* sp. *Regul Toxicol Pharmacol* 2002;**36**:155–61.
6. Salkinoja-Salonen MS, Vuorio R, Andersson MA, Kamfer P, Andersson MC, Honkanen-Buzalski T, et al. Toxigenic strains of *Bacillus licheniformis* related to food poisoning. *Appl Environ Microbiol* 1999;**65**:4637–45.
7. Slepecky RA, Hemphill HE. The prokaryotes. In: Balows A, Trüper HJ, Dworkin M, Harder W, Schleifer KH, editors. *The genus Bacillus—nonmedical*, vol. II. New York: Springer; 1992. p. 1663–96.
8. Chun J, Bae KS. Phylogenetic analysis of *Bacillus subtilis* and related taxa based on partial *gyrA* gene sequences. *Antonie van Leeuwenhoek* 2000;**78**:123–7.
9. Mikkola R, Kolari M, Andersson MA, Helin J, Salkinoja-Salonen MS. Toxic lactonic lipopeptide from food poisoning isolates of *Bacillus licheniformis*. *Eur J Biochem* 2000;**267**: 4068–74.
10. Okstad OA, Gominet M, Purnelle B, Rose M, Lereclus D, Kolsto AB. Sequence analysis of three *Bacillus cereus* loci carrying PlcR-regulated genes encoding degradative enzymes and enterotoxin. *Microbiology* 1999;**145**:3129–38.
11. Beattie SH, Williams AG. Detection of toxigenic strains of *Bacillus cereus* and other *Bacillus* spp. with an improved cytotoxicity assay. *Lett Appl Microbiol* 1999;**28**:221–5.
12. Phelps RJ, McKillip JL. Enterotoxin production in natural isolates of *Bacillaceae* outside the *Bacillus cereus* group. *Appl Environ Microbiol* 2002;**68**:3147–51.
13. Rowan NJ, Caldow G, Gemmel CG, Hunter IS. Production of diarrheal enterotoxins and other potential virulence factors by veterinary isolates of *Bacillus* species associated with nongastrointestinal infections. *Appl Environ Microbiol* 2003;**69**:2372–6.
14. Agata N, Mori M, Ohta M, Suwan S, Ohtani I, Isobe M. A novel dodecadepsipeptide, cereulide, isolated from *Bacillus cereus* causes vacuole formation in HEp-2 cells. *FEMS Microbiol Lett* 1994;**121**:31–4.
15. Agata N, Ohta M, Yokoyama K. Production of *Bacillus cereus* emetic toxin (cereulide) in various foods. *Int J Food Microbiol* 2002;**73**:23–7.
16. Finlay WJ, Logan NA, Sutherland AD. *Bacillus cereus* produces most emetic toxin at lower temperatures. *Lett Appl Microbiol* 2000;**31**:385–9.
17. Häggblom MM, Apetroaie C, Andersson MA, Salkinoja-Salonen MS. Quantitative analysis of cereulide, the emetic toxin of *Bacillus cereus*, produced under various conditions. *Appl Environ Microbiol* 2002;**68**:2479–83.
18. Agata N, Ohta M, Mori M, Isobe M. A novel dodecadepsipeptide, cereulide, is an emetic toxin of *Bacillus cereus*. *FEMS Microbiol Lett* 1995;**129**:17–20.
19. Andersson MA, Mikkola R, Helin J, Andersson MC, Salkinoja-Salonen M. A novel sensitive bioassay for detection of *Bacillus cereus* emetic toxin and related depsipeptide ionophores. *Appl Environ Microbiol* 1998;**64**:1338–43.
20. Lund T, Granum PE. Comparison of biological effect of the two different enterotoxin complexes isolated from three different strains of *Bacillus cereus*. *Microbiology* 1997;**143**: 3329–36.

21. Nordic Committee on Food Analysis. *Bacillus cereus. Determination in foods.* Oslo (Norway): Nordic Committee on Food Analysis No. 67; 1997.
22. Hoult B, Tuxford AF. Toxin production by *Bacillus pumilus. J Clin Pathol* 1991;**44**:455–8.
23. Suominen I, Andersson MA, Andersson MC, Hallaksela MA, Kampfer P, Rainey FA, et al. Toxic *Bacillus pumilus* from indoor air, recycled paper pulp, Norway spruce, food poisoning outbreaks and clinical samples. *Syst Appl Microbiol* 2001;**24**:267–76.
24. Nieminen T, Rintaluoma N, Andersson M, Taimisto AM, Ali-Vehmas T, Seppälä A, et al. Toxinogenic *Bacillus pumilus* and *Bacillus licheniformis* from mastitic milk. *Vet Microbiol* 2007;**124**:329–39.
25. Christiansson A, Naidu AS, Nilsson I, Wadström T, Pettersson H. Toxin production by *Bacillus cereus* dairy isolates in milk at low temperatures. *Appl Environ Microbiol* 1989;**55**:2595–600.
26. Meer RR, Baker J, Bodyfelt FW, Griffiths MW. Psychrotrophic *Bacillus* spp. in fluid milk products - a review. *J Food Prot* 1991;**54**:969–79.
27. Vaisanen OM, Mwaisumo NJ, Salkinoja-Salonen MS. Differentiation of dairy strains of the *Bacillus cereus* group by phage typing, minimum growth temperature, and fatty acid analysis. *J Appl Bacteriol* 1991;**70**:315–24.
28. van Netten P, van De Moosdijk A, van Hoensel P, Mossel DA, Perales I. Psychrotrophic strains of *Bacillus cereus* producing enterotoxin. *J Appl Bacteriol* 1990;**69**:73–9.
29. Stenfors LP, Mayr R, Scherer S, Granum PE. Pathogenic potential of fifty *Bacillus weihenstephanensis* strains. *FEMS Microbiol Lett* 2002;**215**:47–51.
30. Parkinson TJ, Merrall M, Fenwick SG. A case of bovine mastitis caused by *Bacillus cereus. N Z Vet J* 1999;**47**:151–2.
31. Crielly EM, Logan NA, Anderton A. Studies on the *Bacillus* flora of milk and milk products. *J Appl Bacteriol* 1994;**77**:256–63.
32. Jones TO, Turnbull PC. Bovine mastitis caused by *Bacillus cereus. Vet Rec* 1981;**108**: 271–4.
33. Perrin D, Greenfield J, Ward GE. Acute *Bacillus cereus* mastitis in dairy cattle associated with use of a contaminated antibiotic. *Can Vet J* 1976;**17**:244–7.
34. Davies RH, Wray C. Seasonal variations in the isolation of *Salmonella typhimurium, Salmonella enteritidis, Bacillus cereus* and *Clostridium perfringens* from environmental samples. *Zentralbl Veterinarmed B* 1996;**43**:119–27.
35. Scheifer B, Macdonald KR, Klavano GG, van Dreumel AA. Pathology of *Bacillus cereus* mastitis in dairy cows. *Can Vet J* 1976;**17**:239–43.
36. Mavangira V, Angelos JA, Samitz EM, Rowe JD, Byrne BA. Gangrenous mastitis caused by *Bacillus* species in six goats. *J Am Vet Med Assoc* 2013;**242**:836–43.
37. Beecher DJ, Schoeni JL, Wong AC. Enterotoxic activity of hemolysin BL from *Bacillus cereus. Infect Immun* 1995;**63**:4423–8.
38. Graham C. *Bacillus* species and non-spore-forming anaerobes in New Zealand livestock. *Surveillance* 1998;**24**:19.
39. Johnson KG. Bovine mastitis caused by *Bacillus cereus. Vet Rec* 1981;**108**:404–5.
40. Logan NA. *Bacillus* species of medical and veterinary importance. *J Med Microbiol* 1988;**25**:157–65.
41. Hannah Jr WN, Ender PT. Persistent *Bacillus licheniformis* bacteremia associated with an international injection of organic drain cleaner. *Clin Infect Dis* 1999;**29**:659–61.
42. Santini F, Borghetti V, Amalfitano G, Mazzucco A. *Bacillus licheniformis* prosthetic aortic valve endocarditis. *J Clin Microbiol* 1995;**33**:3070–3.
43. Jones BL, Hanson MF, Logan NA. Isolation of *Bacillus licheniformis* from a brain abscess following a penetrating orbital injury. *J Infect* 1992;**24**:103–4.
44. Samples JR, Buettner H. Corneal ulcer caused by a biologic insecticide (*Bacillus thuringiensis*). *Am J Ophthalmol* 1983;**95**(2):258–60.
45. Ryoo NH, Chun HJ, Jeon DS, Kim JR, Park SB. *Bacillus licheniformis* peritonitis in a CAPD patient. *Perit Dial Int* 2001;**21**(1):97.

46. Yuste JR, Franco SE, Sanders C, Cruz S, Fernández-Rivero ME, Mora G. *Bacillus licheniformis* as a cause of a deep skin abscess in a 5-year-old girl: an exceptional case following a plant thorn injury. *J Microbiol Immunol Infect* November 11, 2014. pii: S1684-1182(14)00215-1. http://dx.doi.org/10.1016/j.jmii.2014.08.031. [Epub ahead of print].

47. Credille BC, Woolums AR, Giguère S, Robertson T, Overton MW, Hurley DJ. Prevalence of bacteremia in dairy cattle with acute puerperal metritis. *J Vet Intern Med* 2014;**28**:1606–12.

48. Anburajan L, Meena B, Raghavan RV, Shridhar D, Joseph TC, Vinithkumar NV, et al. Heterologous expression, purification, and phylogenetic analysis of oil-degrading biosurfactant genes from the marine sponge-associated *Bacillus licheniformis* NIOT-06. *Bioprocess Biosyst Eng* February 25, 2015;**38**(6):1009–18. [Epub ahead of print].

49. Chatterjee S. Production and estimation of alkaline protease by immobilized *Bacillus licheniformis* isolated from poultry farm soil of 24 Parganas and its reusability. *J Adv Pharm Technol Res* 2015;**6**(1):2–6.

50. Slimene IB, Tabbene O, Gharbi D, Mnasri B, Schmitter JM, Urdaci MC, et al. Isolation of a Chitinolytic *Bacillus licheniformis* S213 strain exerting a biological control against *Phoma medicaginis* infection. *Appl Biochem Biotechnol* February 11, 2015;**175**(7):3494–506. [Epub ahead of print].

51. Chen XM, Lu HM, Niu XT, Wang GQ, Zhang DM. Enhancement of secondary metabolites from *Bacillus licheniformis* XY-52 on immune response and expression of some immune-related genes in common carp, *Cyprinus carpio*. *Fish Shellfish Immunol* February 19, 2015;**45**(1):124–31. [Epub ahead of print].

52. Chettri R, Tamang JP. *Bacillus* species isolated from tungrymbai and bekang, naturally fermented soybean foods of India. *Int J Food Microbiol* 2015;**197**:72–6.

53. Wu Q, Zhang R, Peng S, Xu Y. Transcriptional characteristics associated with lichenysin biosynthesis in *Bacillus licheniformis* from Chinese Maotai-flavor liquor making. *J Agric Food Chem* 2015;**63**(3):888–93.

54. Caamaño-Antelo S, Fernández-No IC, Böhme K, Ezzat-Alnakip M, Quintela-Baluja M, Barros-Velázquez J, et al. Genetic discrimination of foodborne pathogenic and spoilage *Bacillus* spp. based on three housekeeping genes. *Food Microbiol* 2015;**46**:288–98.

55. Fernández-No IC, Böhme K, Caamaño-Antelo S, Barros-Velázquez J, Calo-Mata P. Identification of single nucleotide polymorphisms (SNPs) in the 16S rRNA gene of foodborne *Bacillus* spp. *Food Microbiol* 2015;**46**:239–45.

56. Paraneeiswaran A, Shukla SK, Prashanth K, Rao TS. Microbial reduction of [Co(III)-EDTA]⁻ by *Bacillus licheniformis* SPB-2 strain isolated from a solar salt pan. *J Hazard Mater* February 2015;**283**:582–90.

57. Laribi-Habchi H, Bouanane-Darenfed A, Drouiche N, Pauss A, Mameri N. Purification, characterization, and molecular cloning of an extracellular chitinase from *Bacillus licheniformis* stain LHH100 isolated from wastewater samples in Algeria. *Int J Biol Macromol* 2015;**72**:1117–28.

58. Wang Z, Wang Y, Zheng L, Yang X, Liu H, Guo J. Isolation and characterization of an antifungal protein from *Bacillus licheniformis* HS10. *Biochem Biophys Res Commun* 2014;**454**(1):48–52.

59. Li T, Deng X, Wang J, Chen Y, He L, Sun Y, et al. Biodegradation of nitrobenzene in a lysogeny broth medium by a novel halophilic bacterium *Bacillus licheniformis*. *Mar Pollut Bull* 2014;**89**(1–2):384–9.

60. Heo J, Kim SK, Park KS, Jung HK, Kwon JG, Jang BI. A double-blind, randomized, active drug comparative, parallel-group, multi-center clinical study to evaluate the safety and efficacy of probiotics (*Bacillus licheniformis*, Zhengchangsheng® capsule) in patients with diarrhea. *Intest Res* 2014;**12**(3):236–44.

61. Andreeva ZI, Nesterenko VF, Yurkov IS, Budarina ZI, Sineva EV, Solonin AS. Purification and cytotoxic properties of *Bacillus cereus* hemolysin II. *Protein Exp Purif* 2006;**47**:186–93.

62. Andreeva ZI, Nesterenko VF, Fomkina MG, Ternovsky VI, Suzina NE, Bakulina AY, et al. The properties of *Bacillus cereus* hemolysin II pores depend on environmental conditions. *Biochim Biophys Acta* 2007;**1768**:253–63.

63. Apetroaie C, Andersson MA, Spröer C, Tsitko I, Shaheen R, Jääskeläinen EL, et al. Cereulide-producing strains of *Bacillus cereus* show diversity. *Arch Microbiol* 2005;**184**:141–51.

64. Baida G, Budarina ZI, Kuzmin NP, Solonin AS. Complete nucleotide sequence and molecular characterization of hemolysin II gene from *Bacillus cereus*. *FEMS Microbiol Lett* 1999;**180**:7–14.

65. Beecher DJ, MacMillan JD. Characterization of the components of hemolysin BL from *Bacillus cereus*. *Infect Immun* 1991;**59**:1778–84.

66. Beecher DJ, Wong AC. Identification and analysis of the antigens detected by two commercial *Bacillus cereus* diarrheal enterotoxin immunoassay kits. *Appl Environ Microbiol* 1994;**60**:4614–6.

67. Beecher DJ, Wong AC. Identification of hemolysin BL-producing *Bacillus cereus* isolates by a discontinuous hemolytic pattern in blood agar. *Appl Environ Microbiol* 1994;**60**:1646–51.

68. Beecher DJ, Wong AC. Improved purification and characterization of hemolysin BL, a hemolytic dermonecrotic vascular permeability factor from *Bacillus cereus*. *Infect Immun* 1994;**62**:980–6.

69. Beecher DJ, Wong AC. Tripartite hemolysin BL from *Bacillus cereus*. Hemolytic analysis of component interactions and a model for its characteristic paradoxical zone phenomenon. *J Biol Chem* 1997;**272**:233–9.

70. Beecher DJ, Wong AC. Cooperative, synergistic and antagonistic haemolytic interactions between haemolysin BL, phosphatidylcholine phospholipase C and sphingomyelinase from *Bacillus cereus*. *Microbiology* 2000;**146**:3033–9.

71. Beecher DJ, Wong AC. Tripartite haemolysin BL: isolation and characterization of two distinct homologous sets of components from a single *Bacillus cereus* isolate. *Microbiology* 2000;**146**:1371–80.

72. Stenfors Arnesen LP, Fagerlund A, Granum PE. From soil to gut: *Bacillus cereus* and its food poisoning toxins. *FEMS Microbiol Rev* 2008;**32**(4):579–606.

73. Zigha A, Rosenfeld E, Schmitt P, Duport C. Anaerobic cells of *Bacillus cereus* F4430/73 respond to low oxidoreduction potential by metabolic readjustments and activation of enterotoxin expression. *Arch Microbiol* 2006;**185**:222–33.

74. Wong HC, Chang MH, Fan JY. Incidence and characterization of *Bacillus cereus* isolates contaminating dairy products. *Appl Environ Microbiol* 1998;**54**:699–702.

75. Yang IC, Shih DY, Huang TP, Huang YP, Wang JY, Pan TM. Establishment of a novel multiplex PCR assay and detection of toxigenic strains of the species in the *Bacillus cereus* group. *J Food Prot* 2005;**68**:2123–30.

76. Yokoyama K, Ito M, Agata N, Isobe M, Shibayama K, Horii T, et al. Pathological effect of synthetic cereulide, an emetic toxin of *Bacillus cereus*, is reversible in mice. *FEMS Immunol Med Microbiol* 1999;**24**:115–20.

77. Vassileva M, Torii K, Oshimoto M, Okamoto A, Agata N, Yamada K, et al. A new phylogenetic cluster of cereulide-producing *Bacillus cereus* strains. *J Clin Microbiol* 2007;**45**:1274–7.

78. Brillard J, Lereclus D. Comparison of cytotoxin cytK promoters from *Bacillus cereus* strain ATCC 14579 and from a *Bacillus cereus* food-poisoning strain. *Microbiology* 2004;**150**:2699–705.

79. Carlin F, Fricker M, Pielaat A, Heisterkamp S, Shaheen R, Salonen MS, et al. Emetic toxin-producing strains of *Bacillus cereus* show distinct characteristics within the *Bacillus cereus* group. *Int J Food Microbiol* 2006;**109**:132–8.

80. Fagerlund A, Ween O, Lund T, Hardy SP, Granum PE. Genetic and functional analysis of the *cytK* family of genes in *Bacillus cereus*. *Microbiology* 2004;**150**:2689–97.

81. Fagerlund A, Brillard J, Fürst R, Guinebretie're MH, Granum PE. Toxin production in a rare and genetically remote cluster of strains of the *Bacillus cereus* group. *BMC Microbiol* 2007;**7**:43.

82. Fagerlund A, Lindbäck T, Storset AK, Granum PE, Hardy SP. *Bacillus cereus* Nhe is a pore-forming toxin with structural and functional properties similar to the ClyA (HlyE, SheA) family of haemolysins, able to induce osmotic lysis in epithelia. *Microbiology* 2008;**154**:693–704.

83. Clavel T, Carlin F, Lairon D, Nguyen-The C, Schmitt P. Survival of *Bacillus cereus* spores and vegetative cells in acid media simulating human stomach. *J Appl Microbiol* 2004;**97**:214–9.

84. Craig CP, Lee WS, Ho MS. *Bacillus cereus* endocarditis in an addict. *Ann Intern Med* 1974;**80**:418–9.

85. Reller LB. Endocarditis caused by *Bacillus subtilis*. *Am J Clin Pathol* 1973;**60**:714–8.

86. Sugar AM, McCloskey RV. *Bacillus licheniformis* sepsis. *JAMA* 1977;**238**:1180–1.

87. Tabbara KF, Tarabay N. *Bacillus licheniformis* corneal ulcer. *Am J Ophthalmol* 1979;**87**:717–9.

88. Balakrishnan I, Baillod RA, Kibbler CC, Gillespie SH. *Bacillus cereus* peritonitis in a patient being treated with continuous ambulatory peritoneal dialysis. *Nephrol Dial Transpl* 1997;**12**:2447–8.

89. Biasioli S, Chiaramonte S, Fabris A, Feriani M, Pisani E, Ronco C, et al. *Bacillus cereus* as agent of peritonitis during peritoneal dialysis (Letter). *Nephron* 1984;**37**:211–2.

90. Al-Hilali N, Nampoory MRN, Johny KV, Chugh TD. *Bacillus cereus* peritonitis in a chronic peritoneal dialysis patient. *Perit Dil Int* 1997;**17**:514–5.

# CHAPTER 12

# Bacillus thuringiensis
# Insecticide Properties

**Vincenzo Savini, Paolo Fazii**

Clinical Microbiology and Virology, Laboratory of Bacteriology and Mycology, Civic Hospital of Pescara, Pescara, Italy

## SUMMARY

The species *Bacillus thuringiensis* is a member of the *Bacillus cereus* group and has become famous in the field of agriculture and biotechnology as its DNA includes insect pathogen genes. These produce insecticidal crystal pore-forming proteins, known as Cry toxins. Cry toxins interact with several insect midgut proteins, thus leading to formation of a prepore oligomer structure and subsequent membrane insertion, with final killing of insect midgut cells by virtue of osmotic shock. *B. thuringiensis* also produces thuringiensin, or β-exotoxin, a thermostable secondary metabolite; it is not a protein but a small-molecule oligosaccharide with insecticidal activity against a wide range of insects, including Diptera, Coleoptera, Lepidoptera, Hymenoptera, Orthoptera, and Isoptera, as well as numerous nematode species.

## BACKGROUND

Control of insects in agriculture as well as of insect vectors of human diseases is mostly achieved by chemical pesticides. Nevertheless, the use of these chemical compounds has led to problems such as environmental pollution, cancer, and immune system disorders.[24] Although microbial insecticides have been proposed as alternatives for chemical ones, their use is still limited, as most microbes display a narrow spectrum of pesticide activity enabling them to kill certain insect species exclusively. Furthermore, they show a low environmental persistence and require precise application practices, given that many such pathogens cannot tolerate irradiation or are specific to young insect larval stages.

The most successful bacterial pest pathogen used for insect control is *B. thuringiensis*, which currently represents about 2% of the total

*The Diverse Faces of Bacillus cereus*
ISBN 978-0-12-801474-5
http://dx.doi.org/10.1016/B978-0-12-801474-5.00012-8

insecticidal market. *B.* is almost exclusively effective against larval stages of various insects and kills them via disruption of the midgut tissue followed by septicemia, which, probably, is caused not only by *B. thuringiensis* itself but also by other bacterial species.[76] *B. thuringiensis* activity relies on insecticidal toxins as well as on an array of further virulence factors contributing to insect killing. The species produces in fact insecticide compounds including Cry proteins, thuringiensin, vegetative insecticidal proteins (VIP), secreted insecticidal protein, zwittermicin A, Mtx-like toxin, and Bin-like toxin.[9,11,25,107–113,134–139]

Upon sporulation, in particular, *B. thuringiensis* forms insecticidal crystal inclusions that are produced by a variety of insecticidal proteins, that is, the above-mentioned Cry toxins. These have a highly selective spectrum of activity, as they may kill a narrow range of pest species. Cry toxins are, specifically, pore-forming toxins secreted as water-soluble proteins that undergo conformational changes in order to insert into the their hosts' membrane, leading to osmotic shock-based killing. Despite the limited use of *B. thuringiensis* products as sprayable insecticides, Cry toxins have been introduced into transgenic crops, thus providing a more targeted and effective way to control agriculture-related pests. Also, such a natural approach has led to a significant reduction in the use of chemical insecticides.[49]

## CRY TOXINS VERSUS LEPIDOPTERAN, DIPTERAN, AND COLEOPTERAN INSECTS

The activity of Cry toxins has mostly been investigated in lepidopteran insects.[11,12,75] Cry toxins are classified based on their primary amino acid sequence and more than 500 different *cry* gene sequences have been studied and categorized into 67 groups (Cry1–Cry67).[13] The *cry* gene sequences have been subdivided into four phylogenetically unrelated protein families showing different modes of action: these are represented by the family of mosquitocidal Cry toxins (Mtx), the family of three-domain Cry toxins (3D), the Cyt family, and the family of binary-like toxins (Bin).[11] Also, certain *B. thuringiensis* strains produce further insecticidal toxins called VIPs that, in contrast to Cry toxins, are formed during the vegetative growth phase, rather than upon sporulation. At least three VIP toxins have been characterized, that is, VIP1/VIP2, a binary toxin, and VIP3.[25,87]

The largest Cry family is the 3D family, which includes at least 40 different groups with more than 200 gene sequences.[8,9,12,34,41,43,55,63] The three-dimensional structure is conserved within the 3D Cry family, suggesting that these proteins share a similar mechanism of action although their amino acid sequence similarity is quite low.[10,20,22,23,86]

Cry toxin insect specificity largely relies on their specific binding to surface proteins of microvilli of larval midgut cells. In the case of lepidopteran insects, Cry1 binding proteins have been characterized as cadherin-like proteins, glycosylphosphatidylinositol (GPI)-anchored aminopeptidase-N (APN), and GPI-anchored alkaline phosphatase (ALP)[35,39,75] and, in general, lepidopteran species including *Manduca sexta, Bombyx mori, Heliothis virescens, Helicoverpa armigera, Pectinophora gossypiella, Ostrinia nubilalis, Lymantria dispar, and Plutella xylostella* are known to be affected by Cry1 molecules.[6,54,61,75]

It is generally accepted that Cry1 toxins may bind cadherin proteins with high affinity in the nanomolar range. Conversely, APN and ALP display lower binding affinities in the range of >100 nM.[2,3,8,17,36,37,51,64,67–69,70,75,77,78,84,89]

As mentioned above, there is general consensus that *B. thuringiensis* Cry toxins act as pore-forming proteins causing cell lysis via osmotic shock. However, an alternative model of the Cry toxin mode of action has been proposed.[94] Specifically, it has been suggested that insect cell death is triggered by the binding of monomeric Cry1Ab toxin to the cadherin receptor, with consequent increases in cAMP cellular levels after activation of adenylyl cyclase. cAMP then activates protein kinase A, thus resulting in cell death.[94] In this model, neither GPI-anchored receptors nor oligomer formation is involved in Cry-related toxicity.[94]

There is an increasing interest in understanding Cry toxins' mode of action in mosquitoes since these insects are important vectors of human diseases, mostly dengue, yellow fever, and malaria. There are multiple Cry toxins showing low primary sequence similarities that display toxicity against mosquitoes, including Cry1, Cry2, Cry4, Cry11, Cry29, and others. Nevertheless, one particular *B. thuringiensis* strain has been used worldwide for mosquito control; this is *B. thuringiensis* var. *israelensis*, which produces crystal inclusions formed principally by Cry4Aa, Cry4Ba, Cry11Aa, and Cyt1Aa.[58] *B. thuringiensis* var. *israelensis* is highly toxic to *Aedes aegypti* (vector of dengue and yellow fevers) and *Culex* species, but moderately toxic against *Anopheles gambiae*, which is known to act as a vector of malaria.[58] It is of further interest that these four *B. thuringiensis* var. *israelensis* toxins show a synergistic effect; particularly, the toxicity displayed by the whole *B. thuringiensis* var. *israelensis* crystal inclusion is much higher than the sum of the individual toxicities shown by each of the Cry and Cyt proteins in this crystal.[11] Cyt1Aa has been observed to synergize the activity of the three Cry toxins and to overcome *Culex* species resistance to Cry toxins.[53,88]

As in lepidopterans, cadherin proteins have been recognized in *A. aegypti* along with *A. gambiae*, in which they show binding to Cry11Aa and Cry4Ba, respectively.[19,46] In *A. aegypti*, cadherin also behaves as a receptor of Cry11Ba toxin from *B. thuringiensis* var. *jegathesan* but shows lower affinity to Cry4Ba protein.[19,56,71] In both *A. aegypti* and *A. gambiae*, cadherin has been observed to be located in the microvilli of the ceca as well as in the microvilli of the posterior gut cells, which are the same sites where Cry11Aa and Cry4Ba bind.[19,46] Also, Both APN and ALP have been identified in *A. gambiae*, *Anopheles quadrimaculatus*, and *A. aegypti* as Cry4Ba-, Cry11Aa-, and Cry11Ba-binding proteins.[1,18,19,27–30,47,56,91,92]

The observation that similar Cry-binding proteins take part in the mechanism of action of Cry toxins in both lepidopteran and dipteran insects seems to suggest that Cry toxins have a conserved mode of action. However, the exact role of the Cry toxin receptors identified in mosquitoes still remains to be understood.[5] One of the most intriguing aspects of *B. thuringiensis* var. *israelensis* toxins, however, is the synergistic effect of Cyt1Aa on the activities of Cry4Aa, Cry4Ba, and Cry11Aa.[16,73,74]

In the field of coleopteran insects, Cry-binding proteins have been identified in *Tenebrio molitor*, *Diabrotica virgifera virgifera*, *Leptinotarsa decemlineata*, and *Anthonomus grandis*.[26,59,65,72] A cadherin protein from *T. molitor* was identified as a Cry3Aa-binding protein, and it was observed to facilitate Cry3Aa oligomer formation.[26] Also, a cadherin protein was recognized as a Cry3Aa receptor in *D. virgifera virgifera*.[65,72] Only one GPI-anchored protein has been identified in coleopteran insects as a putative Cry receptor; this is an ALP from *A. grandis* that is known to bind Cry1B toxin.[59]

Notably, the identification of similar Cry-binding proteins in the three different insect orders and the fact that Cry toxins share a similar three-domain fold suggest that the mechanism of action of these toxins is conserved among lepidopteran, coleopteran, and dipteran insects.

## CRY TOXINS AS BIOINSECTICIDES

Various *B. thuringiensis*-based products have been developed for insect control in agriculture as well as to counter mosquitoes. Most of these are spore-crystal preparations derived from a few wild-type strains including *B. thuringiensis* var. *kurstaki* HD1, *B. thuringiensis* var. *aizawai* HD137, *B. thuringiensis* var. *san diego*, and *B. thuringiensis* var. *tenebrionis*. *B. thuringiensis* var. *kurstaki* products are used to control several leaf-feeding lepidopterans that represent important crop pests or forest pest defoliators.[79] Products from *B. thuringiensis* var. *aizawai* are particularly active against lepidopteran larvae (which are known to feed on stored grains), while those related to *B. thuringiensis* var. *san diego* and *B. thuringiensis* var. *tenebrionis* are effective against beetle pests in agriculture. Finally, as said above, *B. thuringiensis* var. *israelensis* products are important for controlling mosquitoes that behave as vectors of human diseases (including dengue fever and malaria).[79]

*B. thuringiensis*-based sprayable insecticides have limited use in agriculture as Cry toxins are specific to young larval stages, are susceptible to sun radiation, and have a narrow activity against borer insects. Nonetheless, an important breakthrough in the reduction of chemical pesticides in agriculture came with transgenic crops that can express Cry toxins.[49] In such transgenic plants the Cry protein is produced continuously, thus preventing the insecticidal toxin degradation due to UV light, and specifically targets both chewing and boring insects. Interestingly, in 2009, more than 40 million hectares of *B. thuringiensis* crops were grown worldwide, leading to a significant reduction in the use of chemical pesticides and, moreover, contributing in some cases to suppression of certain insects such as *P. gossypiella*, which is a pest of cotton.[49,83]

Nowadays, the major *B. thuringiensis* crops are corn, soya, cotton, and canola.[49] Commercial *B. thuringiensis* cotton expresses the Cry1Ac protein for the control of lepidopterans such as *Helicoverpa zea* and *P. gossypiella*. Again, *B. thuringiensis* cotton expressing Cry1Ac is able to control *H. virescens* and *O. nubilalis*.[21] The next generation of *B. thuringiensis* cotton expresses Cry34Ab/Cry35Ab binary toxin and Cry3Bb (which control coleopteran pests such as *D. virgifera virgifera*) as well as Cry1A, Cry2Ab, and Cry1F (controlling lepidopteran insects including *Spodoptera frugiperda*).[21] Finally, although not commercially available yet, VIP3 has been successfully produced in transgenic corn.[21]

## RESISTANCE TO CRY TOXINS

In the field of *B. thuringiensis* crops the development of insect resistance is of worrisome concern.[15,38,48] Resistance to Cry toxins can be selected for by mutations in the insect pest affecting any of the steps of the mechanism of action of these compounds. For instance, resistance may rely on alterations in Cry toxin activation,[66] toxin sequestration by lipophorin[57] or esterases,[42] enhanced immune response,[44] and alterations in toxin receptors with consequent reduced binding to insect gut membranes.[39,40] Thus far, the major mechanism of toxin resistance in insect pests is represented by the reduction in toxin binding to midgut cells, due to mutations in Cry toxin receptors such as cadherin, ALP, or APN.[32,45,52,62,93] Recently, in *H. virescens*, a resistant allele has been identified to be related to a mutation in a gene coding for an ABC transporter molecule. This mutation affects binding of Cry1A toxin to brush border membrane vesicles.[33]

Insect resistance has been documented in several pests, such as *Plodia interpunctella, P. xylostella, H. zea, S. frugiperda, Busseola fusca,* and *P. gossypiella.*[4,32,50,60,62,80–82,85,90]

Development of insect resistance to Cry toxins in transgenic crops has been delayed by what is known as the "high-dose refuge strategy." This entails planting a remarkable percentage of non-*B. thuringiensis* plants close to *B. thuringiensis* crops expressing a high dose of Cry toxin. Non-*B. thuringiensis* plant refuges are intended to maintain a population of insects that still show susceptibility to Cry toxins. Such susceptible pests mate with resistant individuals that are selected on *B. thuringiensis* plants, thus leading to production of susceptible offspring by virtue of the recessive characteristic of resistant alleles.[82,83] Moreover, it has been shown that the release of sterile *P. gossypiella* females in an eradication program along with the use of *B. thuringiensis* cotton may efficiently slow down the frequency of resistant alleles in the field.[83] This method is an alternative to the refuge strategy with the aim to avoid significant crop losses since non-*B. thuringiensis* plants will suffer damage from insect attack. Also, this strategy may be especially relevant in countries where the refuge strategy is difficult to implement.

Another strategy, finally, includes the use of gene stacking of different Cry toxins with different modes of action in the same plant.[14,31,78]

## DENGUE

Dengue is the fastest advancing vector-borne infectious disease worldwide, and it is estimated that roughly one-third of the world's population lives in dengue-endemic zones. Also, an estimated 50–100 million cases and several hundred thousand cases of severe dengue occur per year.[106]

In the absence of any vaccine or preventive chemotherapies, it is crucial to control the mosquito vectors of the disease (mostly *A. aegypti*), which represent the only way to prevent and control dengue transmission. The flight range of *A. aegypti* is restricted, as these mosquitoes seldom disperse farther than 100 m from the location where they emerge.[96] Both chemical and biological compounds are important components of dengue vector control programs. Unfortunately, however, the widespread use of chemical insecticides has contributed to decreased susceptibility to these agents among *A. aegypti*, particularly in the Americas and the Caribbean.[97,104,106]

Biological control (biocontrol), based on the use of organisms that compete with or otherwise reduce populations of the targeted insect species, is considered an alternative to chemical insecticides in controlling mosquito vectors of disease; in addition, biocontrol offers reduced potential for resistance selection.[106]

In this context, *B. thuringiensis* var. *israelensis*[95] has shown high efficacy against mosquito and black fly larvae.[98,102] *B. thuringiensis* var. *israelensis* exerts its lethal effects by producing several toxic proteins that are ingested by the larvae and are then activated in their gut, where they cause cell membrane disruption and death of the insect. Notably, this action leads to no adverse effects on nontarget invertebrates and vertebrates.[99–101,105] As the mechanism of action is complex and involves more than one protein, the potential for resistance selection is greatly reduced. *B. thuringiensis* var. *israelensis* is available in numerous formulations, which can be applied by hand or as a spray,[98] thus allowing the microorganism to be used in several breeding habitats.

The killing effect of *B. thuringiensis* var. *israelensis* is fast and typically eliminates immature forms from treated containers within 24 h; also, the residual effect ranges between 2 and 4 weeks. This, however, suggests only that *B. thuringiensis* var. *israelensis* is effective in specifically targeted containers receiving treatment, whereas, given the wide number of habitats, its widespread application to all potential breeding sites does not seem to be practical.

While it may be possible to significantly reduce breeding sites within targeted residential areas, treatment must extend beyond these areas, too, to avoid immigration of adult mosquitoes from untreated areas.[96,103]

In summary, there is, then, evidence that *B. thuringiensis* var. *israelensis* is effective in reducing the density of immature vectors of dengue when applied to targeted containers. However, there is limited evidence suggesting that the organism is effective when used as a single agent, in a community setting. Given the increasing insecticide resistance in dengue mosquitoes in several parts of the world, it is crucial to understand that using alternatives to chemical insecticides such as *B. thuringiensis* var. *israelensis* is becoming increasingly important. Nonetheless, there is a clear need for further investigations in this field, to elucidate the efficacy and effectiveness of *B. thuringiensis* var. *israelensis* and to more clearly link entomological outcomes to dengue transmission measures.

## THURINGIENSIN

Thuringiensin, also known as β-exotoxin, is a thermostable exotoxin expressed during vegetative growth and secreted into the supernatant.[114,140–145] It was first identified by McConnell and Richards[115] subsequent to injection of autoclaved supernatant from liquid *B. thuringiensis* culture into numerous species of insect, which led to the insects' death.[115]

Unlike Cry protein, thuringiensin is a nonspecific small-molecule oligosaccharide composed of adenosine, glucose, phosphoric acid, and gluconic diacid. Also, it is thermostable and retains its bioactivity at 121 °C for 15 min.[116,117]

Thuringiensin shows toxicity to insect species belonging to the Diptera, Coleoptera, Lepidoptera, Orthoptera, Hymenoptera, and Isoptera orders, as well as to several nematode species.[118–121] Thuringiensin toxicity in mammals has been the subject of debate for some time. Investigations in this ambit have not clearly established, as of this writing, whether thuringiensin is toxic to humans; nevertheless, it has been banned from public use based on World Health Organization advice.[122–124] However, a number of experiments seem to have shown that the compound is toxic to mammals, as experimental rats inoculated intratracheally developed histologic abnormalities, including disseminated inflammation in both bronchioles and alveoli, bronchial cellular necrosis, and zones of septal thickening with cellular infiltration along with collagen deposits in the alveolar spaces. Thuringiensin is in fact generally considered as a weak activator of adenylate cyclase or an inhibitor of forskolin-stimulated adenylate cyclase; consequently, alterations in pulmonary oxidative–antioxidative status may reasonably play an important role in thuringiensin-induced lung damage.[125–127]

The mechanism behind thuringiensin insecticidal activity is not yet fully understood. Nevertheless, it is known to be an ATP analog interfering with RNA polymerase.[116,128,129] The molecule thus inhibits the synthesis of RNA by virtue of competition with ATP at binding sites, affecting insect molting and pupation and being responsible for teratological effects at sublethal doses.[130–132] The disease caused by thuringiensin in insects is different from that caused by Cry proteins and is observed only during insect molting and pupation. At the minimum effective dose, larvae develop and pupate normally, and only small pupae fail to undergo eclosion, while just a few adults are teratological. Conversely, at higher doses, thuringiensin causes abnormal larval pupation, or the larvae get unresponsive during molting and eventually die.[133]

## FUTURE PERSPECTIVES

Thuringiensin has long been considered to be an adenine nucleotide analog that is able to interfere with RNA polymerase. Also, it may display toxicity toward mammalian cells and has thus been banned from public use.[111,112,116] Nevertheless, it has been shown that thuringiensin is an adenine nucleoside oligosaccharide rather than an adenine nucleotide analog, as formerly thought.[117] These new data about the structure of thuringiensin and its characteristics may inspire further investigation on toxicity mechanisms of this molecule in mammalian cells and related biosafety problems.

B. *thuringiensis* Cry toxins do represent a valuable tool for pest control, thanks especially to the development of Cry toxin-expressing transgenic plants, and this technology has led to a decreased use of chemical insecticides.[49] The selection of resistant insects could, however, be a threat facing this approach. Also, only a few Cry proteins are currently produced in transgenic plants. New Cry proteins that are effective against important pests will be introduced into transgenic crops, thus reducing the potential for selection of insect resistance. Gene stacking in crops will proceed with the introduction of novel Cry genes from novel B. *thuringiensis* isolates or new Cry genes engineered to display improved insecticidal activities. Elucidating the mechanism of action of Cry toxins along with the resistance selection in insects will allow the design of more efficient B. *thuringiensis* crops and spray products, leading to an increasing reduction of the dependence on chemical pesticides and contributing, therefore, to keeping the environment healthy.

## REFERENCES

1. Abdullah MA, Valaitis AP, Dean DH. Identification of a *Bacillus thuringiensis* Cry11Ba toxin-binding aminopeptidase from the mosquito, *Anopheles quadrimaculatus*. *BMC Biochem* 2006;**22**:7–16.
2. Atsumi S, Inoue Y, Ishizaka T, Mizuno E, Yoshizawa Y, Kitami M, et al. Location of the *Bombyx mori* 175 kDa cadherin-like protein-binding site on *Bacillus thuringiensis* Cry1Aa toxin. *FEBS J* 2008;**275**:4913–26.
3. Arenas I, Bravo A, Soberon M, Gomez I. Role of alkaline phosphatase from *Manduca sexta* in the mechanism of action of *Bacillus thuringiensis* Cry1Ab toxin. *J Biol Chem* 2010;**285**:12497–503.
4. Bagla P. Hardy cotton-munching pests are latest blow to GM crops. *Science* 2010;**327**:1439.
5. Bayyareddy K, Andacht TM, Abdullah MA, Adang MJ. Proteomic identification of *Bacillus thuringiensis* subsp. *israelensis* toxin Cry4Ba binding proteins in midgut membranes from *Aedes (Stegomyia) aegypti* Linnaeus (Diptera, Culicidae) larvae. *Insect Biochem Mol Biol* 2009;**39**:279–86.

6. Bel Y, Escriche B. Common genomic structure for the Lepidoptera cadherin-like genes. *Gene* 2006;**381**:71–80.

7. Deleted in review.

8. Boonserm P, Davis P, Ellar DJ, Li J. Crystal structure of the Mosquito-larvicidal toxin Cry4Ba and its biological implications. *J Mol Biol* 2005;**348**:363–82.

9. Boonserm P, Mo M, Ch A, Lescar J. Structure of the functional form of the mosquito larvicidal Cry4Aa toxin from *Bacillus thuringiensis* at a 2.8-Å resolution. *J Bacteriol* 2006;**188**:3391–401.

10. Bravo A. Phylogenetic relationships of *Bacillus thuringiensis* delta-endotoxin family proteins and their functional domains. *J Bacteriol* 1997;**179**:2793–801.

11. Bravo A, Gill SS, Soberón M. *Bacillus thuringiensis* mechanisms and use. In: Gilbert LI, Iatrou K, Gill SS, editors. *Comprehensive molecluar insect science*. Elsevier BV; 2005. . p. 175–206. ISBN: 0-44-451516-X.

12. Bravo A, Gill SS, Soberon M. Mode of action of *Bacillus thuringiensis* Cry and Cyt toxins and their potential for insect control. *Toxicon* 2007;**49**:423–35.

13. Bravo A, Likitvivatanavong S, Gill SS, Soberón M. *Bacillus thuringiensis*: a story of a successful bioinsecticide. *Insect Biochem Mol Biol* 2011;**41**(7):423–31.

14. Bravo A, Soberón M. How to cope with resistance to Bt toxins? *Trends Biotechnol* 2008;**26**:573–9.

15. Cancino-Rodezno A, Alexander C, Villaseñor R, Pacheco S, Porta H, Pauchet Y, et al. The mitogen-activated protein kinase p38 is involved in insect defense against Cry toxins from *Bacillus thuringiensis*. *Insect Biochem Mol Biol* 2010;**40**:58–63.

16. Cantón PE, Reyes EZ, Ruiz I, Bravo A, Soberón M. Binding of *Bacillus thuringiensis* subsp. *israelensis* Cry4Ba to Cyt1Aa has an important role in synergism. *Peptides* 2011;**32**:595–600.

17. Chen J, Hua G, Jurat-Fuentes JL, Abdullah MA, Adang MJ. Synergism of *Bacillus thuringiensis* toxins by a fragment of a toxin-binding cadherin. *Proc Natl Acad Sci USA* 2007;**104**:13901–6.

18. Chen J, Aimanova KG, Pan S, Gill SS. Identification and characterization of *Aedes aegypti* aminopeptidase N as a putative receptor of *Bacillus thuringiensis* Cry11A toxin. *Insect Biochem Mol Biol* 2009;**39**:688–96.

19. Chen J, Aimanova KG, Fernandez LE, Bravo A, Soberón M, Gill SS. *Aedes aegypti* cadherin serves as a putative receptor of the Cry11Aa toxin from *Bacillus thuringiensis* subsp. *israelensis*. *Biochem J* 2009;**424**:191–200.

20. Cohen S, Dym O, Albeck S, Ben-Dov E, Cahan R, Firer M, et al. High-resolution crystal of activated Cyt2Ba monomer from *Bacillus thuringiensis* subs. *israelensis*. *J Mol Biol* 2008;**380**:820–7.

21. Christou P, Capell T, Kohli A, Gatehouse JA, Gatehouse AM. Recent developments and future prospects in insect pest control in transgenic crops. *Trends Plant Sci* 2006;**11**:302–8.

22. de Maagd RA, Weemen-Hendriks M, Stiekema W, Bosch D. Domain III substitution in *Bacillus thuringiensis* delta-endotoxin Cry1C domain III can function as a specific determinant for *Spodoptera exigua* in different, but not all, Cry1-Cry1C hybrids. *Appl Environ Microbiol* 2000;**66**:1559–63.

23. de Maagd RA, Bravo A, Crickmore N. How *Bacillus thuringiensis* has evolved specific toxins to colonize the insect world. *Trends Genet* 2001;**17**:193–9.

24. Devine GJ, Furlong MJ. Insecticide use: contexts and ecological consequences. *Agr Hum Values* 2007;**24**:281–306.

25. Estruch JJ, Warren GW, Mullins MA, Nye GJ, Craig JA, Koziel MG. Vip3A, a novel *Bacillus thuringiensis* vegetative insecticidal protein with a wide spectrum of activities against lepidopteran insects. *Proc Natl Sci USA* 1996;**93**:5389–94.

26. Fabrick J, Oppert C, Lorenzen MD, Morris K, Oppert B, Jurat-Fuentes JL. A novel *Tenebrio molitor* cadherin is a functional receptor for *Bacillus thuringiensis* Cry3Aa toxin. *J Biol Chem* 2009;**284**:18401–10.

27. Fernández LE, Perez C, Segovia L, Rodriguez MH, Gill SS, Bravo A, et al. Cry11Aa toxin from *Bacillus thuringiensis* binds its receptor in *Aedes aegypti* mosquito larvae through loop α-8 of domain II. *FEBS Lett* 2005;**579**:3508–14.

28. Fernández LE, Aimanova KG, Gill SS, Bravo A, Soberón M. A GPI-anchored alkaline phosphatase is a functional midgut receptor of Cry11Aa toxin in *Aedes aegypti* larvae. *Biochem J* 2006;**394**:77–84.

29. Fernández LE, Martinez-Anaya C, Lira E, Chen J, Evans J, Hernández-Martínez S, et al. Cloning and epitope mapping of Cry11Aa-binding sites in the Cry11Aa-receptor alkaline phosphatase from *Aedes aegypti*. *Biochemistry* 2009;**48**:8899–907.

30. Fernández-Luna MT, Lanz-Mendoza H, Gill SS, Bravo A, Soberón M, Miranda-Rios J. An α-amylase is a novel receptor for *Bacillus thuringiensis* subsp. *israelensis* Cry4Ba and Cry11Aa toxins in the malaria vector mosquito *Anopheles albimanus* (Diptera: Culicidae). *Environ Microbiol* 2010;**12**:746–57.

31. Franklin MT, Nieman CL, Janmaat AF, Soberón M, Bravo A, Tabashnik BE, et al. Modified *Bacillus thuringiensis* toxins and a hybrid *B. thuringiensis* strain counter greenhouse-selected resistance in *Trichoplusia ni*. *Appl Environ Microbiol* 2009;**75**:5739–41.

32. Gahan LJ, Gould F, Heckel DG. Identification of a gene associated with Bt resistance in *Heliothis virescens*. *Science* 2001;**293**:857–60.

33. Gahan LJ, Pauchet Y, Vogel H, Heckel DG. An ABC transporter mutation is correlated with insect resistance to *Bacillus thuringiensis* Cry1Ac toxin. *PLoS Genet* 2010;**6**:e1001248.

34. Galitsky N, Cody V, Wojtczak A, Ghosh D, Luft JR, Pangborn W, et al. Structure of the insecticidal bacterial delta-endotoxin Cry3Bb1 of *Bacillus thuringiensis*. *Acta Cryst* 2001 ;**D57**:1101–9.

35. Garczynski SF, Adang MJ. Investigations of *Bacillus thuringiensis* Cry1 toxin receptor structure and function. In: Charles JF, Délécluse A, Nielsen-LeRoux C, editors. *Entomopathogenic bacteria, from laboratory to field application*. Kluwer Academic Publishers; 2000. p. 181–97.

36. Gómez I, Sanchez J, Miranda R, Bravo A, Soberon M. Cadherin-like receptor binding facilitates proteolytic cleavage of helix alpha-1 in domain I and oligomer pre-pore formation of *Bacillus thuringiensis* Cry1Ab toxin. *FEBS Lett* 2002;**513**:242–6.

37. Gómez I, Arenas I, Benitez I, Miranda-Ríos J, Becerril B, Grande G, et al. Specific epitopes of domains II and III of *Bacillus thuringiensis* Cry1Ab toxin involved in the sequential interaction with cadherin and aminopeptidase-N receptors in *Manduca sexta*. *J Biol Chem* 2006;**281**:34032–9.

38. Griffitts JS, Huffman DL, Whitacre JL, Barrows BD, Marroquin LD, Müller R, et al. Resistance to a bacterial toxin is mediated by removal of a conserved glycosylation pathway required for toxine-host interactions. *J Biol Chem* 2003;**278**:45594–602.

39. Griffitts JS, Haslam SM, Yang T, Garczynski SF, Mulloy B, Morris H, et al. Glycolipids as receptors for *Bacillus thuringiensis* crystal toxin. *Science* 2005;**307**:922–5.

40. Griffits J, Aroian RV. Many roads to resistance: how invertebrates adapt to Bt toxins. *BioEssays* 2005;**27**:614–24.

41. Grochulski P, Masson L, Borisova S, Pusztai-Carey M, Schwartz JL, Brousseau R, et al. *Bacillus thuringiensis* CryIA(a) insecticidal toxin: crystal structure and channel formation. *J Mol Biol* 1995;**254**:447–64.

42. Gunning RV, Dang HT, Kemp FC, Nicholson IC, Moores GD. New resistance mechanism in *Helicoverpa armigera* threatens transgenic crops expressing *Bacillus thuringiensis* Cry1Ac toxin. *Appl Environ Microbiol* 2005;**71**:2558–63.

43. Guo S, Ye S, Liu Y, Wei L, Xue J, Wu H, et al. Crystal structure of *Bacillus thuringiensis* Cry8Ea1: an insecticidal toxin toxic to underground pests, the larvae of *Holotrichia parallela*. *J Struct Biol* 2009;**168**:259.

44. Hernández-Martínez P, Navarro-Cerrillo G, Caccia S, de Maagd RA, Moar WJ, Ferré J, et al. Constitutive activation of the midgut response to *Bacillus thuringiensis* in Bt resistant *Spodoptera exigua*. *PLoS One* 2010;**5**:e12795.

45. Herrero S, Gechev T, Bakker PL, Moar WJ, de Maagd RA. *Bacillus thuringiensis* Cry1Ca-resistant *Spodoptera exigua* lacks expression of one of four aminopeptidase N genes. *BMC Genom* 2005;**24**:6–96.

46. Hua G, Zhang R, Abdullah MA, Adang MJ. *Anopheles gambiae* cadherin AgCad1 binds the Cry4Ba toxin of *Bacillus thuringiensis israelensis* and a fragment of AgCad1 synergizes toxicity. *Biochemistry* 2008;**47**:5101–10.

47. Hua G, Zhang R, Bayyareddy K, Adang MJ. *Anopheles gambiae* alkaline phosphatase is a functional receptor of *Bacillus thuringiensis* subsp. *jegathesan* Cry11Ba toxinBacillus thuringiensis jegathesan. *Biochemistry* 2009;**48**:9785–93.

48. Huffman DL, Abrami L, Sasik R, Corbeil J, van der Goot FG, Aroian RV. Mitogen-activated protein kinase pathways defend against bacterial poreforming toxins. *Proc Natl Acad Sci USA* 2004;**101**:10995–1000.

49. James C. *Global status of commercialized Biotech/GM crops: 2009*. ISAAA Brief No. 41. Ithaca (NY): ISAAA; 2009.

50. Janmaat AF, Myers JH. Rapid evolution and the cost of resistance to *Bacillus thuringiensis* in greenhouse populations of cabbage loopers, *Trichoplusia ni*. *Proc R Soc Lond* 2003;**B270**:2263–70.

51. Jiménez-Juárez A, Muñoz-Garay C, Gómez I, Saab-Rincon G, Damian-Alamazo JY, Gill SS, et al. *Bacillus thuringiensis* Cry1Ab mutants affecting oligomer formation are non-toxic to *Manduca sexta* larvae. *J Biol Chem* 2007;**282**:21222–9.

52. Jurat-Fuentes JL, Gahan LJ, Gould FL, Heckel DG, Adang MJ. The HevCaLP protein mediates binding specificity of the Cry1A class of *Bacillus thuringiensis* toxins in *Heliothis virescens*. *Biochemistry* 2004;**43**:14299–305.

53. Khasdan V, Ben-Dov E, Manasherob R, Boussiba S, Zaritsky A. Toxicity and synergism in transgenic *Escherichia coli* expressing four genes from *Bacillus thuringiensis* subsp. *israeliensis*. *Environ Microbiol* 2001;**3**:798–806.

54. Krishnamoorthy M, Jurat-Fuentes JL, McNall RJ, Andacht T, Adang MJ. Identification of novel Cry1Ac binding proteins in midgut membranes from *Heliothis virescens* using proteomic analyses. *Insect Biochem Mol Biol* 2007;**37**:189–201.

55. Li J, Carrol J, Ellar DJ. Crystal structure of insecticidal δ-endotoxin from *Bacillus thuringiensis* at 2.5 Å resolution. *Nature* 1991;**353**:815–21.

56. Likitvivatanavong S, Chen J, Bravo A, Soberón M, Gill SS. Role of cadherin, alkaline phosphatase and aminopeptidase N as receptors of Cry11Ba toxin from *Bacillus thuringiensis* subsp. *jegathesan* in *Aedes aegypti*. *Appl Environ Microbiol* 2010;**77**(1):24–31.

57. Ma G, Roberts H, Sarjan M, Featherstone N, Lahnstein J, Akhurst R, et al. Is the mature endotoxin Cry1Ac from *Bacillus thuringiensis* inactivated by a coagulation reaction in the gut lumen of resistant *Helicoverpa armigera* larvae? *Insect Biochem Mol Biol* 2005;**35**:729–39.

58. Margalith Y, Ben-Dov E. Biological control by *Bacillus thuringiensis* subsp. *israeliensis*. In: Rechcigl JE, Rechcigl NA, editors. *Insect pest management: techniques for environmental protection*. CRC Press; 2000. p. 243.

59. Martins ES, Monnerat RG, Queiroz PR, Dumas VF, Braz SV, de Souza Aguiar RW, et al. Midgut GPI-anchored proteins with alkaline phosphatase activity from the cotton boll weevil (*Anthonomus grandis*) are putative receptors for the Cry1B protein of *Bacillus thuringiensis*. *Insect Biochem Mol Biol* 2010;**40**:138–45.

60. McGaughey WH. Insect resistance to the biological insecticide *Bacillus thuringiensis*. *Science* 1985;**229**:193–5.

61. McNall RJ, Adang MJ. Identification of novel *Bacillus thuringiensis* Cry1Ac binding proteins in *Manduca sexta* midgut through proteomic analysis. *Insect Biochem Mol Biol* 2003;**33**:999–1010.

62. Morin S, Biggs RW, Shriver L, Ellers-Kirk C, Higginson D, Holley D, et al. Three cadherin alleles associated with resistance to *Bacillus thuringiensis* in pink bollworm. *Proc Natl Acad Sci USA* 2003;**100**:5004–9.

63. Morse RJ, Yamamoto T, Stroud RM. Structure of Cry2Aa suggests an unexpected receptor binding epitope. *Structure* 2001;**9**:409–17.

64. Muñoz-Garay C, Portugal L, Pardo-López L, Jiménez-Juárez N, Arenas I, Gómez I, et al. Characterization of the mechanism of action of the genetically modified Cry-1AbMod toxin that is active against Cry1Ab-resistant insects. *Biochim Biophys Acta Biomembr* 2009;**1788**:2229–37.

65. Ochoa-Campuzano C, Real MD, Martínez-Ramírez AC, Bravo A, Rausell C. An ADAM metalloprotease is a Cry3Aa *Bacillus thuringiensis* toxin receptor. *Biochem Biophys Res Commun* 2007;**362**:437–42.

66. Oppert B, Kramer KJ, Beeman RW, Johnson D, McGaughey WH. Proteinase-mediated insect resistance to *Bacillus thuringiensis* toxins. *J Biol Chem* 1997;**272**:23473–6.

67. Ounjai P, Unger VM, Sigworth FJ, Angsuthanasombat C. Two conformational states of the membrane-associated *Bacillus thuringiensis* Cry4Ba deltaendotoxin complex revealed by electron crystallography: implications for toxin-pore formation. *Biochem Biophys Res Commun* 2007;**361**:890–5.

68. Pacheco S, Gómez I, Gill SS, Bravo A, Soberón M. Enhancement of insecticidal activity of *Bacillus thuringiensis* Cry1A toxins by fragments of a toxin-binding cadherin correlates with oligomer formation. *Peptides* 2009;**30**:583–8.

69. Pacheco S, Gomez I, Arenas I, Saab-Rincon G, Rodriguez-Almazan C, Gill SS, et al. Domain II loop 3 of *Bacillus thuringiensis* Cry1Ab toxin is involved in a "ping-pong" binding mechanism with *Manduca sexta* aminopetidase-N and cadherin receptors. *J Biol Chem* 2009;**284**:32750–7.

70. Pardo-López L, Gómez I, Rausell C, Sánchez J, Soberón M, Bravo A. Structural changes of the Cry1Ac oligomeric pre-pore from *Bacillus thuringiensis* induced by N-acetylgalactosamine facilitates toxin membrane insertion. *Biochemistry* 2006;**45**:10329–36.

71. Park Y, Hua G, Abdullah MA, Rahman K, Adang MJ. Cadherin fragments from *Anopheles gambiae* synergize *Bacillus thuringiensis* Cry4Ba's toxicity against *Aedes aegypti* larvae. *Appl Environ Microbiol* 2009;**75**:7280–2.

72. Park Y, Abdullah MA, Taylor MD, Rahman K, Adang MJ. Enhancement of *Bacillus thuringiensis* Cry3Aa and Cry3Bb toxicities to coleopteran larvae by a toxin-binding fragment of an insect cadherin. *Appl Environ Microbiol* 2009;**75**:3086–92.

73. Pérez C, Fernández LE, Sun J, Folch JL, Gill SS, Soberón M, et al. *Bacillus thuringiensis* subsp. *israeliensis* Cyt1Aa synergizes Cry11Aa toxin by functioning as a membrane-bound receptor. *Proc Natl Acad Sci USA* 2005;**102**:18303–8.

74. Pérez C, Muñoz-Garay CC, Portugal L, Sánchez J, Gill SS, Soberón M, et al. *Bacillus thuringiensis* subsp. *israelensis* Cyt1Aa enhances activity of Cry11Aa toxin by facilitating the formation of a pre-pore oligomeric structure. *Cell Microbiol* 2007;**9**:2931–7.

75. Pigott CR, Ellar DJ. Role of receptors in *Bacillus thuringiensis* crystal toxin activity. *Microbiol Mol Biol Rev* 2007;**71**:255–81.

76. Raymond B, Johnston PR, Nielsen-LeRoux C, Lereclus D, Crickmore N. *Bacillus thuringiensis*: an impotent pathogen? *Trends Microbiol* 2010;**18**:189–94.

77. Rodríguez-Almazan CR, Zavala LE, Muñoz-Garay C, Jiménez-Juárez N, Pacheco S, Masson L, et al. Dominant negative mutants of *Bacillus thuringiensis* Cry1Ab toxin function as anti-toxins: demonstration of the role of oligomerization in toxicity. *PLoS One* 2009;**4**:e5545.

78. Soberón M, Pardo-López L, López I, Gómez I, Tabashnik B, Bravo A. Engineering modified Bt toxins to counter insect resistance. *Science* 2007;**318**:1640–2.

79. Soberón M, Gill SS, Bravo A. Signaling versus punching hole: how do *Bacillus thuringiensis* toxins kill insect midgut cells? *Cell Mol Life Sci* 2009;**66**:1337–49.

80. Storer NP, Babcock JM, Schlenz M, Meade T, Thompson GD, Bing JW, et al. Discovery and characterization of field resistance to Bt Maize: *Spodoptera frugiperda* (Lepidoptera: Noctuidae) in Puerto Rico. *J Econ Entomol* 2010;**103**:1031–8.

81. Tabashnik BE. Evolution of resistance to *Bacillus thuringiensis*. *Annu Rev Entomol* 1994;**39**:47–9.
82. Tabashnik BE, Gassman AJ, Crowdwer DW, Carriere Y. Insect resistance to Bt crops: evidence versus theory. *Nat Biotechnol* 2008;**26**:199–202.
83. Tabashnik BE, Sisterson MS, Ellsworth PC, Dennehy TJ, Antilla L, Liesner L, et al. Suppressing resistance to Bt cotton with sterile insect releases. *Nat Biotechnol* 2010;**28**(12): 1304–7.
84. Taveecharoenkool T, Angsuthanasombat C, Kantchanawarin C. Combined molecular dynamics and continuum solvent studies of the pre-pore Cry4Aa trimer suggest its stability in solution and how it may forma pore. *PMC Biophys* 2010;**3**:1–16.
85. van Rensburg JBJ. First report of field resistance by stem borer *Busseola fusca* (Fuller) to Bt-transgenic maize. *S Afr J Plant Soil* 2007;**24**:147–51.
86. Walters FS, deFontes CM, Hart H, Warren GW, Chen JS. Lepidopteran-active variable-region sequence imparts coleopteran activity in eCry3.1Ab, an engineered *Bacillus thuringiensis* hybrid insecticidal protein. *Appl Environ Microbiol* 2010;**76**:3082–8.
87. Warren G. Vegetative insecticidal proteins: novel proteins for control of corn pests. In: Carozzi N, Koziel M, editors. *Advances in insect control: the role of transgenic plants*. Taylor & Francis Ltd; 1997. p. 109.
88. Wirth MC, Georghiou GP, Federeci BA. CytA enables CryIV endotoxins of *Bacillus thuringiensis* to overcome high levels of CryIV resistance in the mosquito, Culex. *Proc Natl Acad Sci USA* 1997;**94**:10536–40.
89. Xie R, Zhuang M, Ross LS, Gómez I, Oltean DI, Bravo A, et al. Single amino acid mutations in the cadherin receptor from *Heliothis virescens* affect its toxin binding ability to Cry1A toxins. *J Biol Chem* 2005;**280**:8416–25.
90. Xu X, Yu L, Wu Y. Disruption of a cadherin gene associated with resistance to Cry1Ac delta-endotoxin of *Bacillus thuringiensis* in *Helicoverpa armigera*. *Appl Environ Microbiol* 2005;**71**:948–54.
91. Zhang R, Hua G, Andacht TM, Adang MJ. A 106-kDa aminopeptidase is a putative receptor for *Bacillus thuringiensis* Cry11Ba toxin in the mosquito *Anopheles gambiae*. *Biochemistry* 2008;**47**:11263–72.
92. Zhang R, Hua G, Urbauer JL, Adang MJ. Synergistic and inhibitory effects of aminopeptidase peptides on *Bacillus thuringiensis* Cry11Ba toxicity in the mosquito *Anopheles gambiae*. *Biochemistry* 2010;**49**(39):8512–9.
93. Zhang S, Cheng H, Gao Y, Wang G, Liang G, Wu K. Mutation of an aminopeptidase N gene is associated with *Helicoverpa armigera* resistance to *Bacillus thuringiensis* Cry1Ac toxin. *Insect Biochem Mol Biol* 2009;**39**:421–9.
94. Zhang X, Candas M, Griko NB, Taussig R, Bulla Jr LA. A mechanism of cell death involving an adenylyl cyclase/PKA signaling pathway is induced by the Cry1Ab toxin of *Bacillus thuringiensis*. *Proc Natl Acad Sci USA* 2006;**103**:9897–902.
95. Goldberg L, Margalit J. A bacterial spore demonstrating rapid larvicidal activity against *Anopheles sergentii, Uranotaenia unguiculata, Aedes aegypti, Culex pipiens, Culex univitattus*. *Mosq News* 1977;**37**:355–8.
96. Harrington C, Scott T, Lerdthusnee K, et al. Dispersal of the dengue vector *Aedes aegypti* within and between rural communities. *Am J Trop Med Hyg* 2005;**72**:209–20.
97. Harris A, Rajatileka S, Ranson H. Pyrethroid resistance in *Aedes aegypti* from Grand Cayman. *Am J Trop Med Hyg* 2010;**83**:277–84.
98. Lacey L. *Bacillus thuringiensis* serovariety *israelensis* and *Bacillus sphaericus* for mosquito control. *J Am Mosquito Control Assoc* 2007;**23**:133–63.
99. Lacey L, Mulla M. Safety of *Bacillus thuringiensis* (H-14) and *Bacillus sphaericus* to non-target organisms in the aquatic environment. In: Laird M, Lacey L, Davidson E, editors. *Safety of microbial insecticides*. Boca Raton: CRC Press; 1990. p. 169–88.

100. Lee B, Scott G. Acute toxicity of temephos, fenoxycarb, diflubenzuron, and metho-prene and *Bacillus thuringiensis* var. *israelensis* to the mummichog (*Fundulus heteroclitus*). *Bull Environ Contam Toxicol* 1989;**43**:827–32.

101. Merritt R, Walker E, Wilzbach M, Cummins K, Morgan W. A broad evaluation of B.t.i. for black fly (Diptera: Simuliidae) control in a Michigan river: efficacy, carry and non-target effects on invertebrates and fish. *J Am Mosquito Control Assoc* 1989;**5**:397–415.

102. Mittal P. Biolarvicides in vector control: challenges and prospects. *J Vector Borne Dis* 2003;**40**:20–32.

103. Muir L, Kay B. *Aedes aegypti* survival and dispersal estimated by mark–release–recapture in northern Australia. *Am J Trop Med Hyg* 1998;**58**:277–82.

104. Rodriguez M, Bisset J, De Armas Y, Ramos F. Pyrethroid insecticide-resistant strain of *Aedes aegypti* from Cuba induced by deltamethrin selection. *J Am Mosquito Control Assoc* 2005;**21**:437–45.

105. Saik J, Lacey L, Lacey C. Safety of microbial control agents to domestic animals and vertebrate wildlife. In: Laird M, Lacey L, Davidson E, editors. *Safety of microbial insecticides*. Boca Raton: CRC Press; 1990. p. 115–31.

106. Boyce R, Lenhart A, Kroeger A, Velayudhan R, Roberts B, Horstick O. *Bacillus thuringiensis israelensis* (Bti) for the control of dengue vectors: systematic literature review. *Trop Med Int Health* 2013;**18**(5):564–77.

107. Schnepf E, Crickmore N, van Rie J, Lereclus D, Baum J, Feitelson J, et al. *Bacillus thuringiensis* and its pesticidal crystal proteins. *Microbiol Mol Biol Rev* 1998;**62**:775–806.

108. Sudakin DL. Biopesticides. *Toxicol Rev* 2003;**22**:83–90.

109. Ye W, Zhu L, Liu Y, Crickmore N, Peng D, Ruan L, et al. Mining new crystal protein genes from *Bacillus thuringiensis* on the basis of mixed plasmid-enriched genome sequencing and a computational pipeline. *Appl Environ Microbiol* 2012;**78**:4795–801.

110. Donovan WP, Engleman JT, Donovan JC, Baum JA, Bunkers GJ, Chi DJ, et al. Discovery and characterization of Sip1A: a novel secreted protein from *Bacillus thuringiensis* with activity against coleopteran larvae. *Appl Microbiol Biotechnol* 2006;**72**:713–9.

111. Stabb EV, Jacobson LM, Handelsman J. Zwittermicin A-producing strains of *Bacillus cereus* from diverse soils. *Appl Environ Microbiol* 1994;**60**:4404–12.

112. Crickmore N, Zeigler DR, Feitelson J, Schnepf E, van Rie J, Lereclus D, et al. Revision of the nomenclature for the *Bacillus thuringiensis* pesticidal crystal proteins. *Microbiol Mol Biol Rev* 1998;**62**:807–13.

113. Wirth MC, Walton WE, Federici BA. Inheritance patterns, dominance, stability, and allelism of insecticide resistance and cross-resistance in two colonies of *Culex quinquefasciatus* (Diptera: Culicidae) selected with cry toxins from *Bacillus thuringiensis* subsp. *israelensis*. *J Med Entomol* 2010;**47**:814–22.

114. Sharma CB, Prasad SS, Pai SB, Sharma S. The exotoxin of *Bacillus thuringiensis*: a new C-mitotic agent. *Experientia* 1976;**32**:1465–6.

115. McConnell E, Richards AG. The production by *Bacillus thuringiensis* Berliner of a heat stable substance toxic for insects. *Can J Microbiol* 1959;**5**:161–8.

116. Farkas J, Sebesta K, Horska K, Samek Z, Dollijs J, Storm F. The structure of exotoxin of *Bacillus thuringiensis* var. *gelechiae*. *Collect Czechslov Chem Commun* 1969;**34**:1118–20.

117. Liu XY, Ruan LF, Hu ZF, Peng DH, Cao SY, Yu ZN, et al. Genome wide screening reveals the genetic determinants of an antibiotic insecticide in *Bacillus thuringiensis*. *J Biol Chem* 2010;**285**:39191–200.

118. Toledo J, Liedo P, Williams T, Ibarra J. Toxicity of *Bacillus thuringiensis* beta-exotoxin to three species of fruit flies (*Diptera: Tephritidae*). *J Econ Entomol* 1999;**92**:1052–6.

119. Tamez-Guerra P, Iracheta MM, Pereyra-Alférez B, Galán-Wong LJ, Gomez-Flores R, Tamez-Guerra RS, et al. Characterization of Mexican *Bacillus thuringiensis* strains toxic for lepidopteran and coleopteran larvae. *J Invertebr Pathol* 2004;**86**:7–18.

120. Tsuchiya S, Kasaishi Y, Harada H, Ichimatsu T, Saitoh H, Mizuki E, et al. Assessment of the efficacy of Japanese *Bacillus thuringiensis* isolates against the cigarette beetle, *Lasioderma serricorne* (Coleoptera: Anobiidae). *J Invertebr Pathol* 2002;**81**:122–6.

121. Zhang JH, Wang CZ. Comparison of toxicity and deterrence among crystal, spore and thuringiensin A of *Bacillus thuringiensis* against *Helicoverpa armigera* (Hubner). *Acta Entomol Sin* 2000;**43**:85–91.

122. World Health Organization. *Guidelines specification for bacterial larvicides for public health use*. Geneva, Switzerland: World Health Organization; 1999. Publication WHO/CDS/CPC/WHOPES/99.2. Report of the WHO Informal Consultation.

123. Meretoja T, Garlberg G. The effect of *Bacillus thuringiensis* and of cell-free supernatants of some other bacteria on the mitotic activity of human lymphocytes. *FEMS Microbiol* 1977;**2**:109–11.

124. Barjac H, de Dedoder R. Purification and complementary analysis of the thermostable toxin of *Bacillus thuringiensis* var. *thuringiensis*. *Bull Soc Chim Biol* 1968;**50**:941–4.

125. Tsai SF, Liu BL, Liao JW, Wang JS, Hwang JS, Wang SC, et al. Pulmonary toxicity of thuringiensin administered intratracheally in Sprague-Dawley rats. *Toxicology* 2003;**186**:205–16.

126. Tsai SF, Yang C, Wang SC, Wang JS, Hwang JS, Ho SP. Effect of thuringiensin on adenylate cyclase in rat cerebral cortex. *Toxicol Appl Pharmacol* 2004;**194**:34–40.

127. Tsai SF, Yang C, Liu BL, Hwang JS, Ho SP. Role of oxidative stress in thuringiensin-induced pulmonary toxicity. *Toxicol Appl Pharmacol* 2006;**216**:347–53.

128. Beebee T, Korner A, Bond RP. Differential inhibition of mammalian ribonucleic acid polymerases by an exotoxin from *Bacillus thuringiensis*. The direct observation of nucleoplasmic ribonucleic acid polymerase activity in intact nuclei. *Biochem J* 1972;**127**:619–34.

129. Sebesta K, Horska K. Mechanism of inhibition of DNA-dependent RNA polymerase by exotoxin of *Bacillus thuringiensis*. *Biochim Biophys Acta* 1970;**209**:357–76.

130. Burgerjon A, Biache G, Cals P. Teratology of the Colorado potato beetle, *Leptinotarsa decemlineata*, was provoked by larval administration of the thermostable toxin of *Bacillus thuringiensis*. *J Invertebr Pathol* 1969;**14**:274–8.

131. Ignoffo CM, Gregory BG. Effects of *B. thuringiensis* beta-exotoxin on larval maturation, adult longevity, fecundity, and egg viability in several species of lepidoptera. *Environ Entomol* 1972;**1**:269–72.

132. Espinasse S, Gohar M, Chaufaux J, Buisson C, Perchat S, Sanchis V. Correspondence of high levels of beta-exotoxin I and the presence of *cry1B* in *Bacillus thuringiensis*. *Appl Environ Microbiol* 2002;**68**:4182–6.

133. Royalty RN, Hall HR, Lucius BA. Effects of thuringiensin on *Tetranychus urticae* (Acari: Tetranychidae) mortality, oviposition rate and feeding. *Pestic Sci* 1991;**33**:383–91.

134. Barjac H, Burgerjon A, Bonnefoi A. The production of heat-stable toxin by nine serotypes of *Bacillus thuringiensis*. *J Invertebr Pathol* 1966;**8**:537–8.

135. Kim YT, Huang HT. The β-exotoxin of *Bacillus thuringiensis* isolation and characterization. *J Invertebr Pathol* 1970;**15**:100–8.

136. Tsun HY, Liu CM, Tzeng YM. Recovery and purification of thuringiensin from the fermentation broth of *Bacillus thuringiensis*. *Bioseparation* 1999;**7**:309–16.

137. Gohar M, Perchat S. Sample preparation for β-exotoxin determination in *Bacillus thuringiensis* cultures by reversed-phase high-performance liquid chromatography. *Anal Biochem* 2001;**298**:112–7.

138. Levinson BL, Kasyan KJ, Chiu SS, Currier TC, Gonzalez JM. Identification of beta-exotoxin production, plasmids encoding beta-exotoxin, and a new exotoxin in *Bacillus thuringiensis* by using high-performance liquid chromatography. *J Bact* 1990;**172**:3172–9.

139. Bekheit HK, Lucas AD, Gee SJ. Development of an enzyme linked immunosorbent assay for the β-exotoxin of *Bacillus thuringiensis*. *J Agric Food Chem* 1993;**41**:1530–4.
140. Liu CM, Tzeng YM. Quantitative analysis of thuringiensin by HPLC using AMP as internal standard. *J Chromatogr Sci* 1998;**36**:340–4.
141. Campbell DP, Dieball DE, Brackett JM. Rapid HPLC assay for the beta-exotoxin of *Bacillus thuringiensis*. *J Agric Food Chem* 1987;**35**:156–8.
142. De Rijk TC, van Dam RC, Zomer P, Boers EA, de Waard P, Mol HG. Development and validation of a confirmative LC-MS/MS method for the determination of β-exotoxin thuringiensin in plant protection products and selected greenhouse crops. *Anal Bioanal Chem* 2013;**405**:1631–9.
143. Perani M, Bishop AH, Vaid A. Prevalence of β-exotoxin, diarrhoeal toxin and specific δ-endotoxin in natural isolates of *Bacillus thuringiensis*. *FEMS Microbiol Lett* 1998;**160**: 55–60.
144. Espinasse S, Gohar M, Lereclus D, Sanchis V. An ABC transporter from *Bacillus thuringiensis* is essential for beta-exotoxin I production. *J Bacteriol* 2002;**184**:5848–54.
145. Espinasse S, Gohar M, Lereclus D, Sanchis V. An extracytoplasmic-function sigma factor is involved in a pathway controlling beta-exotoxin I production in *Bacillus thuringiensis* subsp. *thuringiensis* strain 407-1. *J Bacteriol* 2004;**186**:3108–16.

# INDEX

Printed in the United States
By Bookmasters